Advance Praise for *The Science of Mom*

"Yes! An easy-to-read, fascinating, nuanced review of the science behind new parents' biggest health questions. Many of these issues—infant sleep, breastfeeding, vaccines—have or will hit your 'Should I panic?' button. With gentle guidance, Alice Callahan puts your fears to rest."

—TRACY CUTCHLOW, author of *Zero to Five: 70 Essential Parenting Tips Based on Science (and What I've Learned So Far)*

"Too many of today's parents treat science as a weapon, using it to justify some choices and condemn others. Yet, most don't fully understand what science can and cannot tell us. By giving parents a comprehensive understanding of how science relates to parenting, Alice Callahan has helped us turn this weapon into a tool for peace. Callahan untangles basic scientific concepts, reveals the realities and limitations of research, and advocates for a measured approach to parenting science that eschews absolutes and acknowledges nuance. *The Science of Mom* is a rare gem in the parenting canon--smart, sensitive, and a lifesaver for a generation of parents caught in the nebulous spider's web of Internet 'wisdom.'"

—SUZANNE BARSTON, author of *Bottled Up: How the Way We Feed Babies Has Come to Define Motherhood, and Why It Shouldn't*

"Families routinely search for health information. *The Science of Mom* makes it easy collecting evidence for health decisions and putting it into perspective with a mom-to-mom connection. Callahan's advice is thoughtful, backed by science, and feels fueled by love. She is willing to provide powerful advice when detailing the science and safety of vaccines. Keep this book in arm's reach as you support your infant for calm and direction."

—WENDY SUE SWANSON, MD, MBE, FAAP
Seattle Children's Hospital, author of *Mama Doc Medicine: Finding Calm and Confidence in Parenting, Child Health, and Work-Life Balance*

"Fascinating! Think of all the controversial, hot-button topics that parents obsess about in a child's first year—from vaccines and feeding to bed-sharing and sleep training. Weighing the scientific evidence, Callahan offers balanced insights and in-depth answers—a far cry from the oversimplified advice prescribed by many 'parenting experts.' The result: a must-have guide that's substantive and extremely engaging."

—JENA PINCOTT, author of *Do Chocolate Lovers Have Sweeter Babies? The Surprising Science of Pregnancy*

The Science of Mom

The *Science* of *Mom*

A Research-Based Guide to Your Baby's First Year

ALICE CALLAHAN, PhD

JOHNS HOPKINS UNIVERSITY PRESS * *Baltimore*

 Published 2015
Printed in the United States of America on acid-free paper
9 8 7 6 5 4 3 2

Johns Hopkins University Press
2715 North Charles Street
Baltimore, Maryland 21218-4363
www.press.jhu.edu

Library of Congress Cataloging-in-Publication Data

Callahan, Alice, 1980–
The science of mom : a research-based guide to your baby's first year /
Alice Callahan, PhD
pages cm
Includes bibliographical references and index.
ISBN 978-1-4214-1732-5 (pbk. : alk. paper) — ISBN 978-1-4214-1733-2 (electronic) —
ISBN 1-4214-1732-4 (pbk. : alk. paper) — ISBN 1-4214-1733-2 (electronic)
1. Infants—Care. 2. Infants—Health and hygiene.
I. Title.
RJ61.C23 2015
618.92'01—dc23——2014041138

A catalog record for this book is available from the British Library.

Special discounts are available for bulk purchases of this book. For more information, please contact Special Sales at 410-516-6936 or specialsales@press.jhu.edu.

Johns Hopkins University Press uses environmentally friendly book materials, including recycled text paper that is composed of at least 30 percent post-consumer waste, whenever possible.

FOR MY TWO CHARLOTTES

the one who brought me
into the world

and the one who brought me
into motherhood

CONTENTS

ACKNOWLEDGMENTS

Before I became a parent, I had no idea how much work it would be, how much I had to learn, and how radically it would change my life. I can say the same about writing this, my first book. And just like raising a child, writing a book, I now know, takes lots of help. I couldn't have done it on my own.

This book would never have happened without the gracious support of the readers of my blog, *Science of Mom*. They convinced me that there are other parents, like me, who want and need evidence-based information on parenting. They inspired me to get started, encouraged me to keep going, and were patient when my blog grew quiet so that I could focus on finishing the book.

Writing this book gave me a good reason to fully immerse myself in lots of fun and interesting parenting science. I'm grateful to all the scientists who have trained and mentored me over the years. They probably had no idea how I might apply this skill set, but I think I've put it to good use.

Most of the science in this book came from my digging through the published literature, but conversations with researchers and clinicians also helped clarify and illuminate the data for me and sometimes revealed fascinating stories behind the science. For this, my thanks to Drs. Thomas Anders, Helen Ball, Peter Blair, Martin Blaser, Camila Chaparro, Leita Dzubay, Jeffrey Ecker, Penny Glass, Frank Greer, Margaret Hammerschlag, Douglas Leonard, Lauren

Marcewicz, Jodi Mindell, Rachel Moon, Paul Offit, Tonse Raju, Henry Redel, Robert Sidonio, Paul Slovic, Douglas Teti, and Kristi Watterberg. Just as much, I am grateful to all of the parents who generously shared their real life stories, helping me to put the science in context: Suzanne Barston, Eve F., Jordan and Cheryl Green, Margaret Green, Stefani Leavitt, Esmee McKee, Janie Oyakawa, Leah R., and Valerie Wheat.

Thank you to my editor, Vincent Burke, at Johns Hopkins University Press for first recognizing the potential in this book, and to him and the staff at the Press for expertly guiding me through the publishing process.

I am forever indebted to my mom, Charlotte Green, for many reasons. She was my first example of gentle and respectful parenting, and she raised me to be a reader and a writer. She was also my best editor throughout the process of writing this book. She let my preoccupation with the book take over our long weekly phone conversations, and she thoughtfully read and critiqued chapter after chapter.

Many other friends and family members also read chapters and provided valuable feedback: Rob Callahan, Sarah Holexa, Dorit Reiss, Sarah Ruttan, Jessica Smock, Miya Tokumitsu, and Robin White. Thanks also to the anonymous peer reviewer who provided honest and helpful comments on the manuscript.

Thank you to Rob, my husband, friend, and partner in parenting, for his constant encouragement and many fun daddy days with Cee while I worked away on the book. And finally, thank you to Cee, for being my inspiration, keeping me humble, and growing my heart.

The Science of Mom

INTRODUCTION

I was a scientist first. I started working in research labs as a college student, then as a lab technician, and then went on to graduate school for a PhD in nutritional biology. By the time my husband and I decided we were ready for a baby, I was a postdoctoral fellow studying fetal physiology. For the previous decade of my life, I had been working to understand the world through the lens of science. I asked questions and sought answers by designing and conducting experiments in the lab. I pored over journal articles, trying to make sense of my data and figure out what important questions to ask next. My job was to measure, analyze, and explain. If I felt lost, then the way out of darkness was always to learn more, maybe by repeating an experiment or immersing myself in the scientific literature until things started to look a little clearer.

It should come as no surprise, then, that when I didn't get pregnant in the first couple of months of trying, I tackled it in the best way I knew how: with science. I measured my basal body temperature and tracked my cervical mucus. I recorded my data in a spreadsheet and searched for patterns in color-coded graphs. I pulled out my reproductive physiology textbook to brush up on ovulation and how to optimize our chances at fertilization. It's hard to say whether any of this helped me to get pregnant any faster, but it did at least give me a feeling of control, though perhaps at the expense of romance.

When I eventually did get pregnant and became a mother, this same story played out again and again, with different details. Pregnancy and childbirth were just the beginning. Caring for my daughter, Cee, brought lots more questions. Where should she sleep, and when could I expect her to sleep through the night? When should she start solid foods, and what should I feed her? Should I be worried about vaccine safety? Between feeding and rocking, changing and bouncing, and soothing and singing, I found myself digging into the science of parenting.

At about the same time, I left the research lab so that I could spend more time with Cee and explore other career paths, but I missed talking about science with like-minded colleagues. I started the *Science of Mom* blog as a place to share and discuss what I was learning about parenting, and I was thrilled to find a community of equally inquisitive parents looking for solid science. My blog readers were fascinated by the complexities of the science and always ready with new questions, which inspired me to keep researching and writing. At some point, I wanted to be able to develop these topics in a more cohesive manner, and that journey became this book.

Why do we look to science for answers to parenting questions? For most of human history, how we cared for our babies was passed quietly from generation to generation, one mother to the next. It varied around the world, and it changed over time, but it was informed by culture, environment, and necessity—not science. But for our generation, new parenthood means making a huge number of decisions with little familial or cultural guidance. Most of us live far away from the parents and grandparents, aunts and uncles, and brothers and sisters who might have guided us into parenthood in generations past. Instead, we have unlimited advice from the Internet, a resource that empowers us to find our own answers but at the same time overwhelms us with confusing information. If our babies won't nap or we're struggling with breastfeeding, we can ask Google for answers or solicit opinions from social networks like Facebook and Twitter. Very quickly, we find ourselves in a minefield of conflicting information and opinions, often paired with a little judgment from all sides.

Science is a tool that can help us get closer to the truth in the midst of all this confusion. It can help us to sidestep the opinions and philosophies and anecdotes that fuel parenting debates. We look to science, not to confirm what we want to believe or to prove others wrong, but rather to make the best decisions we can as parents.

It isn't that science has all the answers to our parenting questions. For some questions, like those that are medical in nature, science can be a great guide. For others, like parenting practices, it can often only give us clues that we then have to figure out how to apply to our own families. The truth is, from a scientific standpoint, parenting is an immensely complicated thing to study. How do we sort through and quantify so many variables—along with all the differences between babies and families—and, statistically, make some sense of it all? Something so complex can rarely be reduced to a right way and a wrong way. What science can do is zoom out to look at large numbers of babies and families to reveal patterns and averages and ranges of normal. It can give us a broader perspective that can be hard to find during the day-to-day tasks of caring for a baby. It can reveal risks and benefits that we may not have considered, allowing us to approach decisions in an objective way. Science may not offer a protocol for parenting, like the procedures that I perfected in my research lab, but it can help us make smart decisions for our families.

Science can also help us rise above parenting controversies. In many cases, if you carefully study the data, as I did in my research for this book, you see that there's rarely a strong case for everyone doing things in the same way. There's plenty of room for parents to make different choices about how their baby eats and sleeps, for example. When you understand the limitations of the science, you are empowered to follow your heart and your child's lead. And in cases where the science gives us a clearer answer, it can help us make decisions more confidently. We don't have to waste our energy arguing or defending ourselves, because we've looked at the evidence, and we're comfortable with our choices. This frees up valuable time and energy to focus on our babies, take a nap, or work on things important to us instead of scouring conflicting Internet articles for nuggets of truth.

I begin this book, in chapter 1, with a sort of crash course in evidence-based parenting. If we want to examine parenting decisions in a scientific way, then we have to understand how science works, how to parse through many studies to find the most relevant ones, and how to interpret research in a meaningful way. This outline will give you an idea of how I approached each topic in this book, and it should also prepare you to tackle your own questions with science as your guide.

In chapters 2 and 3, I zoom in on a few medical questions important to newborn babies, including when we should cut the umbilical cord and why newborns receive an injection of vitamin K and a smear of antibiotics in their eyes. These may be small questions in the scheme of a baby's first year of life, but they each provide a fascinating case study in the history and science of newborn medicine. In chapter 4, I step away from decision making and appreciate new babies as scientists do, for their incredible abilities to sense and explore their new world with touch, sound, sight, and smell. Understanding our newborns sets the stage for how we care for them responsively and how the parent-child relationship develops over time.

In the rest of the book, I examine some of the biggest questions and sources of angst for parents: breastfeeding and bottle feeding (chapter 5); sleep safety and bed sharing (chapter 6); baby sleep patterns (chapter 7); vaccine safety (chapter 8); starting solid foods (chapter 9); and healthy nutrition for the older baby (chapter 10). As I researched and wrote each chapter of the book, I read hundreds of scientific papers, talked to scientists, and interviewed parents for their real life stories. I looked for answers, but I also wasn't afraid to investigate and question the parts of science where things are still uncertain, and that is often where I found some of the most interesting stories.

I didn't write this book pretending to be a parenting expert. I am just a mom with big questions, and I had the scientific training and the curiosity to dig for the answers. I didn't set out to argue one side or the other on controversial topics. Instead, I've tried to be honest about what the science does and doesn't tell us, even when that might be different from the simplified recommendations handed down to us from parenting authority figures. I loved getting buried in convoluted literature searches, trying to find some kind of truth that could be useful to you and me. Thanks for coming along on this journey with me. I hope it makes your job as a parent a little easier—and maybe more interesting, too.

1 ⁂ SHOW ME THE SCIENCE

A Crash Course in Evidence-Based Parenting

Until I became pregnant, my science life and my real life rarely converged. I spent long hours in the laboratory and in reading and writing scientific papers. But for the most part, when I went home, I left the science behind. I saw the scientific process as a way for me to understand my little corner of research, not as a helpful tool in making everyday decisions.

All of that changed when I became pregnant. Suddenly I had a lot of questions and a lot of decisions to make, and these felt really important. I realized that, as a parent, I was tasked with the responsibility of making decisions for another person, and I wanted to get this right. I had to figure out what chemicals and procedures in my research lab might pose a health risk to my baby. I wanted to eat right and exercise well for an optimal pregnancy. I wanted to plan and prepare for the birth in the same detail-oriented way that I mapped out a protocol in the lab. And science, I realized, was a tool that could help me do this, and best of all, I knew how to use it. Good data comforted me. As a scientist, surely I could make this an evidence-based pregnancy and an evidence-based baby.

Science did help me to make some decisions, particularly in how I planned for childbirth and parenting. But I also found that the reality often deviated from the plan. No amount of planning could prepare me for how a labor contraction would feel as it screamed through my body, or for how I would cope when my baby had been up for hours, needing *me*, in the middle of the night. I learned that part of parenting is accepting that it usually won't go as planned and that we can't control every variable. We have to experience it for ourselves and learn as we go. We adjust and adapt and then go back to the science with new questions, wanting to learn more.

Many of the areas of research that I investigated for this book were much more complex than any scientific question I had tackled before. I was accustomed to cell culture or animal studies, where I could control nearly every variable, tweaking the one in question to determine its importance or to get my experimental conditions just right. But parenting questions have so many more variables. For example, most of us, at some point, wonder how we can help our babies to sleep better at night. But how can we answer that with science? We can't put babies in a laboratory and control every variable that might affect their sleep. Babies are born to different parents, who have different philosophies and cultural traditions that they carry with them into parenting. Babies live in different types of environments with varying amounts of light, noise, and activity around them. Even if we could control all of those things, the babies themselves are different, born with their own unique personalities and needs.

All of this natural variation between babies and between families makes parenting research difficult to do and even harder to interpret. Taking an evidence-based approach to parenting means being aware of these limitations and approaching the science with skepticism. Science can be such a valuable tool, but only if we understand how it works. We have to be able to find relevant studies, judge their quality, interpret their results with skepticism, and then figure out whether they can inform our own approaches to parenting.

In this book, I explore what I think are some of the most interesting parenting decisions in a baby's first year, but it isn't a complete guide by any means. This book probably won't answer all of your questions, but I think you'll find that many of the concepts introduced here will help you to tackle other questions on your own. To help you understand how I approached my research for this book and how you can do your own research, let's walk through some of the most important concepts for understanding science.

1. SCIENCE IS A PROCESS.

The scientific method allows us to observe the world in a systematic, objective way. Children learn the scientific method, starting in elementary school, and I have fond memories of conducting my science fair experiments and preparing my presentation with Question, Hypothesis, Prediction, Experimental Method, Data, and Conclusions in big, stenciled letters. This seems like a pretty simple system, and it is. It is exactly this simplicity and its systematic approach that makes it so valuable. It means that other scientists can repeat the experiment to verify it or not, and still others might change one or two variables to determine how they affect the results. What really makes it meaningful is that we all use the same process, and over time, it helps us to better understand the world. Good science takes lots of experiments, time, and people to build a body of knowledge. But it always starts in the same place: with a question, and the desire to answer it.

2. GOOD SCIENCE IS PEER-REVIEWED.

A critical part of the scientific process is peer review and publication of results. Someone may be a brilliant scientist, conducting groundbreaking research in her lab, but if she doesn't publish her results, then they are essentially meaningless. Her data will die when she dies. Publication is a permanent record of scientific discovery. When studies are published, other researchers can check the work and build on it.

If a scientist wants to publish her research, she has to write it up in a research paper and submit it to a scientific journal. Journals, especially the best ones, receive many more papers than they actually publish, and they're selective about what they accept. If the editor thinks a paper could be a good fit for the journal, it will be subjected to peer review. This means that experts (usually three or four) in that specific area of research will read and critique the paper. They'll check to see whether the researcher used good methodology and statistics to test her hypothesis and whether her interpretations of the results are reasonable. Peer reviewers aren't paid to do this job, but they take the responsibility seriously, knowing that the legitimacy of their field of research rests on good peer review. They're also usually anonymous, so they can freely offer their opinions and concerns without fear of damaging relationships with other scientists. The peer review system is a filter for bad science; flawed studies won't get published (although this means they are often resubmitted to a lower quality journal). Peer reviewers often ask authors to collect more data, analyze it in a different way, or consider alternative explanations for the results. The purpose of peer review is to keep everyone honest and to keep the standards of scientific publication high.

If you're looking to science for answers to your parenting questions, be sure you're looking at peer-reviewed science. Most scientific journals are peer-reviewed, but you can check journal web pages to be sure. Letters to the editor and commentaries published in journals aren't usually peer-reviewed.

3. ONE STUDY IS NEVER ALL THAT MEANINGFUL ON ITS OWN; WHAT MATTERS IS SCIENTIFIC CONSENSUS.

Scientific knowledge has to start somewhere, but one study is never worth much on its own, even after peer review. Scientists can make mistakes, use flawed methodology, or simply misinterpret their data, so we like to see studies repeated and hypotheses tested in different ways before we get too excited about results.

Andrew Wakefield's paper attempting to link the measles, mumps, and rubella (MMR) vaccine to autism is a dramatic example of this.[1] Wakefield's study used methodology that was weak and ethically questionable, and it didn't actually provide evidence that the MMR shot caused autism. Somehow, it got through peer review and was published in the *Lancet*, a prestigious journal. But since scientists are never satisfied with one study, more researchers studied this question, looking at it from different angles and using better methods, and their results didn't support Wakefield's findings. We now have a huge amount of science showing that Wakefield was wrong, and his paper was retracted, essentially removing it from the archives of reputable science (more on that story in appendix D). So, while the system of science isn't foolproof, it is usually self-correcting, over time.

One study can't tell us much on its own, but scientific consensus means we're getting closer to really understanding something. If you put a bunch of scientists together in a room, what do 95% of them agree on? If only a couple of studies have been conducted on the topic, you're likely to hear a lot of debate and disagreement in the room. Scientists are skeptical by nature, after all. But over time, as different scientists using different methodologies conduct more and more studies, a consensus emerges. For example, there is a strong scientific consensus that evolution explains the diversity of life, that global warming is happening because of human activity, and that the benefits of vaccines far outweigh the risks. Our understanding of these ideas will continue to grow and evolve, but they're unlikely to change radically, because they are already backed by so many studies.

Other areas of science have less certainty. Some examples discussed in this book are the safety of bed sharing (chapter 6) and the right time to start solid foods (chapter 9). These questions are hotly debated in scientific communities. That's a sign to us that maybe the data are not so solid and there are some aspects of these topics that science hasn't explained very well yet. Researchers' arguing about these questions drives the field forward and urges them to conduct better studies. And it tells us that there's room within evidence-based parenting to make different choices.

4. SOME STUDIES ARE MORE VALUABLE THAN OTHERS.

The great thing about the Internet is that we all have scientific databases at our fingertips. You can search PubMed or Google Scholar for your topic of interest and have access to millions of scientific abstracts (summaries of research papers), and in some cases you can even access the full text. (Unfortunately, unless you are affiliated with an academic institution, many journals will require you to pay a hefty fee for a copy of the article.)

Access to millions of articles is useful only if you can make sense of them. Let's say you're wondering when you should introduce solid foods to your baby, so you search PubMed for "infant solid foods introduction." You get 392 results. You know that most of these articles have passed a certain degree of scrutiny already, since they are published in peer-reviewed journals. But is each of these 392 articles equally important? How do you figure out which are most relevant and helpful? Important factors to consider are the type of article and the study design. Here's a quick guide, listed roughly in order from most to least relevant for answering your parenting and health questions.[2]

* *Systematic reviews and meta-analyses:* A *systematic review* is a survey of studies published on a specific topic or question. Its purpose is to try to summarize what the research says, collectively, about the topic. The authors systematically search scientific databases for studies of the topic, and then they whittle down their results to include the studies that are most relevant and of highest quality, using predetermined criteria. Following their methods, if you had access to the same scientific databases, you should be able to end up with the same collection of studies. The authors of the systematic review report on the overall findings, giving you a summary of the field, including where results are conflicting. This type of paper is useful because it interprets a lot of studies and puts them in context with each other. You will also find review papers that aren't systematic, but these are less reliable because the authors may have cherry-picked studies that they liked and left others out.

A *meta-analysis* is usually a systematic review that goes one step further, combining the quantitative data from each of the relevant studies and conducting a statistical analysis on these combined data. A well-known and reliable source for systematic reviews and meta-analyses is the Cochrane Collaboration, an international nonprofit organization that asks experts to conduct reviews, using very rigorous methods, for the purpose of informing evidence-based medicine. If you're looking for information about when to start solid foods, for example, you're in luck because there is a Cochrane Review on a closely related topic, the optimal duration of exclusive breastfeeding (comparing four versus six months).[3]

* *Randomized controlled trials (RCTs):* An RCT is the highest quality design for a clinical trial. In an RCT, researchers recruit a group of people who first agree to participate in the study and are then randomly allocated to two or more groups, one of which is a control group. For example, an RCT published in 2013 investigated the question of when to start solid foods.[4] A group of exclusively breastfeeding moms in Iceland were randomized to receive advice to start solid foods at either 4 months or 6 months, and then their babies were observed through infancy and into childhood. Because they were randomly assigned, these two groups should have been very similar in every way except for the time when the babies started solid foods. If the babies that started sol-

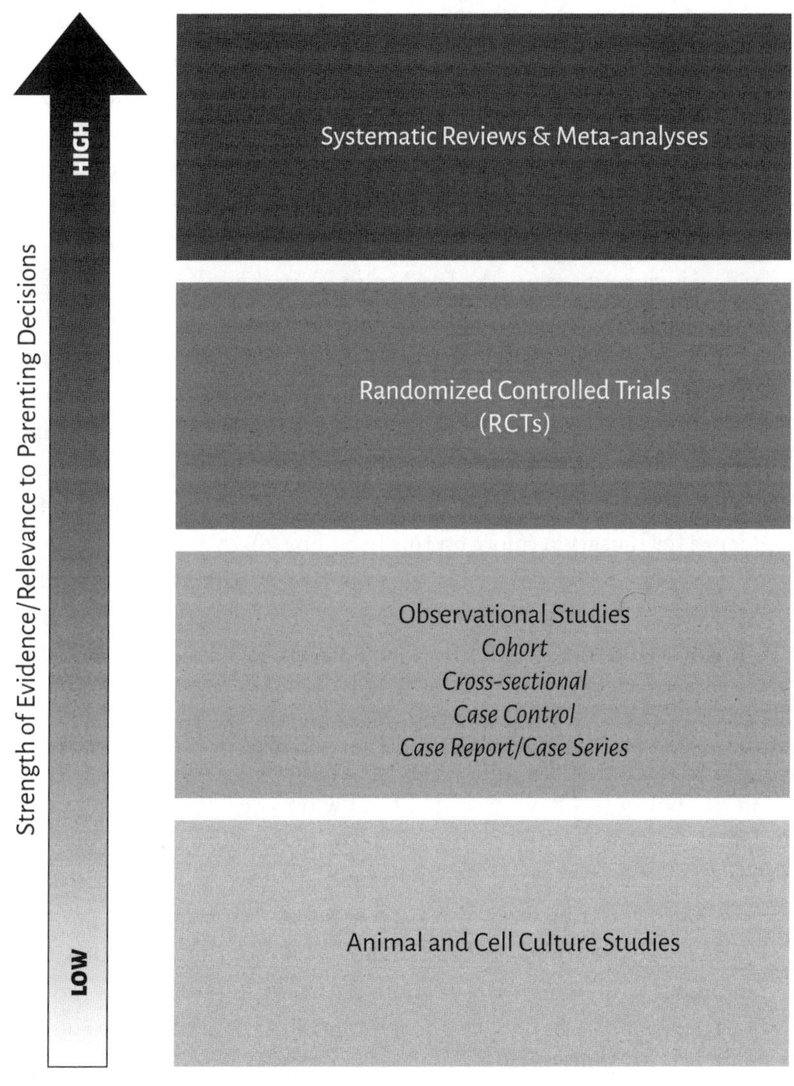

A rough guide to scientific study designs and their relevance to parenting decisions, ranked, from top to bottom, from most useful to least useful.

id foods at 4 months had different measures of later growth or health than the babies that started solids at 6 months, then we would feel fairly certain that the timing of solids *caused* those differences. RCTs are really valuable, but they are also difficult and often unethical to do in parenting research. Most parents don't want a scientist to dictate to them how they should raise their baby, and if we already have data showing that one way is better than another, it would be unethical to tell some parents to use the less healthful method.

* *Observational human studies:* Most parenting studies are observational. On the question of when to start solid foods, I have found only three RCTs conducted to date but hundreds of observational studies, including several categories described below. The most important thing to understand about all of these types of observational studies is that they can only show correlations; they can't provide evidence for causation (more on this in a minute).

A *cross-sectional study* of this topic might look at a bunch of babies at the same age—maybe 12 months—providing a snapshot comparison of health and growth in babies who started solid foods earlier or later. A *cohort study* would observe these babies longitudinally, over time, giving us more information about how variables interact and change as the babies grow. Observational studies can only tell us so much, however, because we don't know how the babies that started solids early or late (or their parents) may have been different from the outset. For example, maybe the babies whose parents gave them solids at 4 months were also bigger to start with or were growing more quickly all along, and maybe they were more interested in solids and more developmentally ready to try them earlier on. This growth pattern could affect both the decision of when to start solids and the outcomes later in the baby's life. It's really hard to tease out these variables, called *confounding factors*.

Another type of observational study is a *case-control study*, in which subjects are identified that have something in common, usually a medical condition or outcome, and are compared with controls that don't have that condition. Case-control studies are discussed in detail in chapter 6, on sleep safety and SIDS.

The least useful type of observational study is a *case report*, a report of a single medical occurrence. For example, a pediatrician might write a case report on a baby who was allergic to rice cereal. This tells you that this allergy can occur and how it appeared in that one baby—important information—but it doesn't tell you anything about how common it is or what might increase or decrease the risk of it happening. A *case series* is simply a group of similar case reports published together. Case reports and case series usually have little relevance when it comes to our parenting decisions.

⁎ *Animal and cell culture studies:* A lot of important science starts in petri dishes containing cell cultures, is then tested in animals, and if it looks promising, may be tested in humans. These are important studies because they can help us to understand mechanisms for how things work, but they are of little use in answering parenting questions. Human babies aren't mice, and they're much more complex than cells grown in a petri dish.

Another type of paper that you might encounter is a policy statement or committee opinion. These can be excellent summaries of the evidence, and they're important because they guide clinical practice for health care workers. However, they aren't usually systematic reviews, and they can be affected by authors' bias. For example, the American Academy of Pediatrics (AAP) Section on Breastfeeding and the AAP Committee on Nutrition disagree on when babies should start solid foods, even though, presumably, they look at the same studies.

One more point: for health questions, it's best to stick to studies that have been conducted in developed nations. Patterns of disease and health outcomes are very different in countries where families don't have a clean source of water or refrigeration for food storage, for example.

5. NUMBERS MATTER.

Studies that include more subjects are more useful. It's hard to give a rule of thumb here, but I'll try. For most of the parenting questions in this book, the best studies should include hundreds of babies. For some of the biggies, like breastfeeding, we find studies with hundreds to thousands of babies. For something like vaccine safety, we expect studies to have thousands to tens of thousands of children. In reality, scientists usually work to ensure that their study includes an adequate number of subjects to test their hypothesis, but if you're comparing two studies of the same question with a similar design, the one with the larger sample size is usually more useful.

6. DON'T BELIEVE EVERYTHING YOU READ ON THE INTERNET.

You can use the tools I just described to sift through the research literature, but this is a ton of work. In my research for this book, I read hundreds of studies, trying to judge their quality and draw meaningful conclusions from them, but this took me a lot of time. Meanwhile, I had my own questions as my daughter turned from baby to toddler to preschooler, and because I was working on this book, I had to find quicker answers. Like most of us, I turned to the Internet, hoping that someone else could summarize the research for me. However, I have high standards for websites that I'll trust when it comes to parenting or health information.

I consider the sites from educational or governmental institutions (ending in .edu or .gov) to be more reliable, usually, than sites from for-profit or nonprofit organizations (.com or .org). I look for sites that are current, that aren't trying to sell anything, and are written by someone who has some formal training in the field and/or cite peer-reviewed studies to back up their claims. I check at least some of those studies to see whether they really support what the website author is saying. If I think I have found a reliable source, then I look for another one to see whether it says something similar. I steer clear of sites with sensationalist claims, personal attacks, and conspiracy theories.

7. CORRELATION IS NOT CAUSATION.

This is so important. If your baby stayed up an hour past her bedtime and then took her first unassisted steps the next day, would you infer that slight sleep deprivation is good for gross motor development? Of course not. There is a correlation between these two events—staying up late and achieving a new milestone—but you can recognize that this is most likely a coincidence.

The same caution is needed when looking at large data sets. One excellent example is the strong correlation between organic food consumption

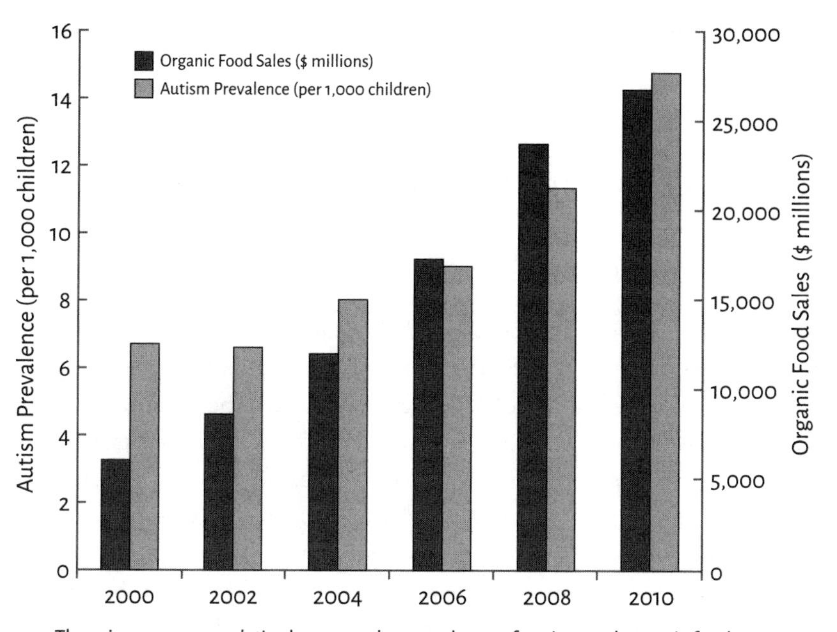

There is a strong *correlation* between the prevalence of autism and organic food sales, but it is unlikely that organic food *causes* autism (or vice versa). *Sources:* Organic Trade Association, "2011 Organic Industry Overview," 2011, http://www.ota.com/pics/documents/2011OrganicIndustrySurvey.pdf; Centers for Disease Control and Prevention, "Autism Spectrum Disorder: Data and Statistics," March 24, 2014, http://www.CDC.gov/ncbddd/autism/data.html.

and autism diagnoses.[5] Would you conclude from the graph above that organic food causes autism? Or that autism makes people buy more organic food? Neither is a likely explanation for the data in the graph. What is more likely is that the two variables are completely unrelated to each other. (There is also good evidence that the main explanation for the increasing diagnosis of autism is increased awareness and recognition, not a true epidemic of the condition.)[6]

This is not to say that correlations are never useful. Sometimes a correlation is the beginning of an important finding. For example, case-control studies in the 1980s showed a correlation between SIDS and tummy sleeping. There wasn't really evidence that tummy sleeping *caused* SIDS, but this was a consistent finding in multiple studies, and health organizations started recommending that babies be put "Back to Sleep" to prevent SIDS. This campaign was followed by a dramatic drop in SIDS rates.[7] Later studies also showed that babies sleeping on their tummies sleep more deeply and have a harder time arousing from sleep, providing a potential mechanism, or explanation, for tummy sleeping causing SIDS.[8] While tummy sleeping and SIDS started out as just a correlation, we now feel pretty certain that this relationship also represents causation.

8. WE CAN'T ELIMINATE RISK.

Everything we do carries some risk. When we feed our children, we risk choking, food allergies, and food-borne pathogens. When we put them down to sleep, there is a risk of SIDS. And yet, clearly, the advantages of eating and sleeping outweigh these small risks. Our job, as parents, is to do our best to minimize the risks within the context of our daily lives, but this means considering all sources of risk, some of which may be personal to us and not considered in the scientific studies or the policy statements.

Take breastfeeding, for example. We know that formula-fed babies, on average, have a higher risk of infection than breastfed babies. All else being equal, most parents would want to minimize the risk of infection for their babies. But what about the mom who, to safely breastfeed, has to stop taking the medication that normally helps her manage her mental illness? In this situation, there is also a risk to breastfeeding that can jeopardize the mother's health and her relationship with her baby. Risks and benefits are personal, and while science can give us some understanding of risks and benefits in a population of study subjects, we also have to trust that each family does its best to balance the risks and benefits that are personal to that family.

9. FIND SMART ALLIES.

I consider myself pretty savvy about science, but I constantly get paralyzed by the need to make decisions, in part because I really care about making the right ones. I've come to accept that I can't know everything, and sometimes I need help to make evidence-based choices. For this reason, I work to find health care providers that are evidence-based, that I trust, and that are willing to look at the science with me and help me make good decisions. I expect them to listen to my concerns, but then I listen to their advice and value their training and expertise. A good pediatrician can be a tremendous resource. I also recommend looking for good sources of outside information that can help compile evidence for you. Find blogs (like *Science of Mom!*) that make this their mission, and follow the work of science and health journalists that do a good job of reporting on parenting issues.

10. FORGET ABOUT PERFECTION, AND PAY ATTENTION TO *YOUR* BABY.

Parenting brings so many decisions, and approaching them from an evidence-based perspective seems to require a crash course in epidemiology, immunology, nutrition, medicine, physiology, and psychology. It's easy to be misled and take a wrong turn, but keeping an open, curious, and skeptical mind will help keep you on track with understanding the science. Even so, no amount of science will mean that you'll make the right decisions all the time. So much of parenting is trial and error; it's part of how we learn about our children and ourselves. The most important work we do as parents can't be described by science, or at least not very well. It's in our daily, mundane interactions with our kids. It's in the way we show them we care for them and the way we give them space to become their own people. It's in the way we talk to them, read to them, play with them, and let them explore the world. These are the things that make our kids who they are, not to mention each child's innate temperament and personality. Use science to the extent that it helps you make choices with confidence or to the extent that it just plain fascinates you, but recognize that it is just one of many tools in your parenting kit.

2 ⁂ CUTTING THE UMBILICAL CORD

When Is the Right Time?

I mark my transition to motherhood as the roughly 30 hours that I was in labor with Cee: starting at midnight on a Sunday and ending on a sunny Tuesday morning. It was a surreal, wild time, marked not by the clock but by contractions. Once I started pushing, there was no measure of time but the nurse's counting me through pushes, over and over and over, for several hours, until finally, I pushed my baby girl into the world.

And just like that, she was here. At first, she was a little blue and limp, but within a minute or so, she took her first breath. She was making her own transition, in which she had to quickly adapt to a vastly different world. With that first breath, her lungs, which had been compressed throughout her fetal life, expanded with air, and blood flowed into an intricate network of pulmonary vessels.

Before birth, Cee had relied on the placenta as her source of life and building blocks for her growth and development. Our placenta was the interface between my blood and her blood. Here, our two blood supplies sidled up next to each other, not mixing, but passing close enough that substances could exchange across a few layers of cells. The umbilical cord was Cee's lifeline to the placenta. Through the cord and across the placenta, nutrients and oxygen passed from me to Cee, and carbon dioxide and waste products were handed off from Cee to me.

To a new baby, part of a healthy birth transition means making the switch from getting oxygen from the placenta to getting oxygen from air, through the lungs. But this switch doesn't have to be abrupt. At birth, whether vaginal or by cesarean, the umbilical cord is still attached to the placenta, which will be delivered shortly after the baby. ("You're not done yet, Alice," I remember my obstetrician reminding me.) Even as the baby breathes, circulation continues to and from the placenta, through the umbilical cord.

If the cord is left intact, it will naturally close and blood will stop flowing within a few minutes. But over the course of those minutes, much of the blood in the placenta will flow toward the baby.[1] Cut the cord immediately, and that blood stays with the placenta, which is usually thrown out with the rest of the hospital's waste. Waiting to cut the cord for two to three minutes after a vaginal birth can give 75 to 100 ml (milliliters) of additional blood (about one-third of a cup) to the average full-term newborn, increasing the baby's total blood volume by up to 50%.[2] This short delay must have happened throughout most of human history, and it is certainly what occurs after the birth of most mammals. Cord clamping can also be delayed in a cesarean birth, although the amount of transfusion from the placenta may be somewhat less than in a vaginal birth.[3]

When to cut the cord has been a topic of discussion for at least the past few centuries.[4] Erasmus Darwin, physician, and grandfather to Charles Darwin, wrote in his 1796 medical text, *Zoonomia*: "Another thing very injurious to the child, is the tying and cutting [of] the navel-string too soon; which should always be left till the child has not only repeatedly breathed, but till all pulsation in the cord ceases. As otherwise the child is much weaker than it ought to be; a part of the blood being left in the placenta, which ought to have been in the child."[5]

Despite grandfather Darwin's admonition, there was a shift in the early to mid-1900s to cutting the cord within 10 to 15 seconds of birth.[6] This practice is now very common, and not just in Western developed nations. Surveys of practices around the world—including countries such as Bangladesh, China, Egypt, Ethiopia, and Tanzania—reveal that most babies now have their cord cut immediately.[7] This change in routine procedure probably came with the shift of childbirth from the realm of midwives in the home to that of doctors in the hospital, where efficiency became a goal and the umbilical cord was, quite literally, the dividing line between the specialties of obstetrics and pediatrics. But this was a practice of convenience; there was never any evidence that cutting the cord immediately is beneficial. Over the past few decades, timing of cord clamping has again come under scrutiny, with researchers wondering how this practice affects the baby.

In this chapter, we take a look at the decision of when to clamp the umbilical cord. It's one of the first things we grown-ups do to a new baby after birth, and timing may matter more than was previously thought. And although this is just one small decision among many that we make as parents, this story is a fascinating case study in the history and science behind childbirth choices.

AN INSIDE LOOK AT CORD CLAMPING RESEARCH

In the early 2000s, Camila Chaparro was a young graduate student of nutrition at the University of California, Davis, working under Dr. Kay Dewey. I was a grad student in the same program around the same time, but our experiences were vastly different. I spent most of my grad school years working in a tedious research lab on campus. Camila spent several years in a large obstetrics hospital in Mexico City, conducting a randomized controlled trial of early versus delayed cord clamping.

In her mid-twenties, having seen only a video of childbirth, Camila found herself signing up study participants in the hospital's large labor rooms. As Camila put it to me, "I knew basically nothing,"[8] and this was an interesting place to begin an education in childbirth. There were usually around six women laboring in one room, with no family members allowed. There was no privacy for these moms; in addition to Camila and her study staff, the labor rooms were crowded with obstetricians, anesthesiologists, and medical and nursing students.

For each woman that agreed to be included in the study, Camila or one of her assistants randomly drew a sealed envelope that assigned the laboring mom to either early or delayed cord clamping. This is what made the study a randomized controlled trial, the highest quality study design. Because the two groups were randomly assigned, the only thing that should have been different about them was the timing of cord clamping. As the baby was born, Camila or an assistant would stand with a stopwatch and give the obstetrician the okay to cut the cord after 10 seconds for the early group or after 2 minutes for the late group. They then took a range of measurements in the babies, beginning on the first day of life and following up until they were 6 months old, to see whether timing of cord clamping made a difference to their health.

Camila and her colleagues found that timing did matter a great deal, and they published their findings in the prestigious medical journal *Lancet*.[9] It was one of the most careful studies in this field, and it was followed by studies from other researchers confirming the benefits of delayed cord clamping.

WHAT A DIFFERENCE TWO MINUTES MAKES

Delaying cord clamping gives a baby more blood at birth, carrying with it approximately 40 to 75 mg (milligrams) of additional iron.[10] Iron is an essential part of the protein hemoglobin, found in red blood cells and responsible for transporting oxygen around the body. In the first weeks after birth, old red blood cells break down, and the iron is recycled and stored in the body, to be used as needed by the baby in the coming months.[11] Iron stores are like a nutritional savings account that the baby will spend down as he grows and develops. In addition to its role in hemoglobin, iron is also needed for muscle growth, as well as brain function and development. It's required for the synthesis of neurotransmitters like serotonin and dopamine and for the myelination of nerve tissue, which allows rapid delivery of nerve signals around the body.[12]

With delayed cord clamping, a baby starts life with more iron, and this comes in handy several months after birth. This was evident in Camila's study. At 6 months of age, the babies with delayed cord clamping had 88% more iron stored than the babies whose cords had been clamped immediately.[13] Other studies have confirmed that babies with delayed cord clamping are less likely to be anemic at 2 to 3 months of age and have higher iron stores at 4 to 6 months of age.[14]

So what happens if a baby becomes iron-deficient? If severe, iron deficiency results in low blood hemoglobin and is one of several causes of anemia. Iron-deficiency anemia during early childhood is associated with lower scores on cognitive, motor, and behavioral tests.[15] In children older than 2 years of age, giving iron supplements to correct an iron deficiency usually results in improved test scores.[16] However, children who had iron-deficiency anemia at 6 months of age and then received an iron supplement to correct the deficiency still had slower activation of nerve pathways as toddlers and preschoolers,[17] suggesting that damage during the early years may be irreversible.[18] (Correcting a deficiency is still important to prevent further damage, however.) Moderate iron deficiency without anemia, which is more common in developed countries, is also associated with cognitive and motor deficits, though the results are subtler and less conclusive.[19]

Iron stores are especially critical for breastfed babies, because there is little iron in breast milk. (Formula is fortified with iron, so iron deficiency is unusual for formula-fed babies.) Breast milk iron is well absorbed, but the amount is so small that it can provide only a fraction of the baby's needs. A breastfed baby thus counts on the iron stores from birth to provide most of what is needed for growth and development in infancy.[20] How long those stores last depends on how big they were at birth. With early cord clamping, iron stores can dwindle by 3 to 4 months of age. Late cord clamping adds several months' worth of iron, making it last until 6 to 8 months.[21] Other factors, including maternal anemia, diabetes, and premature birth, can also reduce the iron stores present at birth and increase the risk of iron deficiency.[22]

The iron bump from delayed cord clamping is vital, because most babies aren't ready to eat an appreciable amount of solid foods until around 6 months of age, and some take longer to get the hang of solids. To cover this gap in iron supply, the AAP's Committee on Nutrition recommends that, beginning at 4 months, breastfed babies take a liquid iron supplement.[23] But this advice is controversial. The AAP's own Section on Breastfeeding responded to this recommendation with major concerns and, along with other researchers in this field, worried that supplementing all breastfed infants could be risky.[24] For example, one study of Swedish and Honduran infants found that iron supplementation was helpful to babies that had low iron, but for babies that already had sufficient iron, the supplement slowed growth and increased the incidence of diarrhea.[25] There's also the very real problem that iron supplements literally taste like rust, and some babies flat-out reject them. I tried giving one to Cee for a while, but she invariably spat it back at me, staining whatever clothes we were wearing. Delayed cord clamping is a physiologically normal way to improve babies' iron stores, and it takes only a few minutes at the time of birth.

In the United States, approximately 11% of 1-year-olds are iron-deficient, and 2% of toddlers have iron-deficiency anemia.[26] These rates are much lower than in developing countries, but still, we're talking about hundreds

✲ SUMMARY OF BENEFITS AND RISKS OF DELAYED CORD CLAMPING

BENEFITS	
In preterm infants	Higher blood pressure, blood volume, hematocrit, and hemoglobin in the newborn period Reduced need for blood transfusions Reduced incidence of intraventricular hemorrhage Reduced incidence of necrotizing enterocolitis Improved cardiac function Improved oxygenation of the brain
In full-term infants	Higher hematocrit and hemoglobin in the newborn period Higher iron stores and reduced risk of iron deficiency and anemia in infancy Reduced risk of lead poisoning
RISKS	
In full-term infants	Possible increased risk of jaundice in the newborn period

of thousands of U.S. children affected by iron deficiency. And importantly, babies in the United States aren't routinely screened for anemia (by measuring hemoglobin concentration) until around 12 months of age, and most are never screened for iron deficiency without anemia.[27] In other words, an older baby, particularly an exclusively breastfed baby whose cord was clamped early and who isn't yet getting good dietary sources of iron, could easily have a moderate iron deficiency go unnoticed in a critical period of development. This is where delayed cord clamping can be so helpful; it gives a baby a few extra months of stored iron—just enough to bridge the transition from exclusive breastfeeding to starting solid foods. (I discuss good options for iron-rich solid foods in chapter 10.)

An added benefit of delayed cord clamping is that it may protect babies from lead poisoning, because iron deficiency increases lead absorption. In another report by Camila Chaparro and her coauthors, Mexican breastfed infants with delayed cord clamping had lower blood lead levels.[28] The

Centers for Disease Control and Prevention (CDC) estimates that four million U.S. households have children exposed to lead, so this benefit has the potential to be highly relevant to children in both developed and developing countries.[29]

BENEFITS OF DELAYED CORD CLAMPING IN PREMATURE BABIES

Delayed cord clamping is helpful to full-term infants, but it is most beneficial for preterm infants. Waiting to cut the cord reduces anemia and the need for blood transfusions in preemies, who are already at greater risk for iron deficiency.[30] Most importantly, delayed cord clamping reduces the incidence of intraventricular hemorrhage (bleeding in the brain) and of necrotizing enterocolitis (death of intestinal tissue).[31] Both of these conditions are life threatening and are common complications of preterm birth. Delayed cord clamping has also been shown to improve oxygenation of brain tissue in premature babies,[32] and in another study, it appeared to improve motor development in 7-month-old boys who were born prematurely.[33] Many of the studies of preterm babies used only a 30- to 45-second delay in cord clamping because of the rush to get medical attention to these fragile babies, but benefits were observed even with this short delay.

RISKS OF DELAYED CORD CLAMPING

One of the most trusted institutions in evidence-based medicine is the Cochrane Collaboration, which publishes systematic reviews of studies of medical interventions. Cochrane asks researchers in the field to prepare a review based on strict criteria for study quality in order to draw conclusions from the best data available. The most recent Cochrane review on delayed cord clamping in full-term babies reports a host of benefits, as just discussed, but it notes one risk: increased jaundice.[34]

Jaundice is caused by a buildup of bilirubin, a normal breakdown product of red blood cells. All newborns are breaking down red blood cells, but in some babies, the liver can't filter bilirubin quickly enough. If bilirubin levels get too high or jaundice doesn't resolve on its own, this can be treated with phototherapy—placing the baby under special lights that help metabolize bilirubin. Without treatment, bilirubin can damage the brain, so monitoring and timely treatment are important. It makes sense that babies with delayed cord clamping could be at higher risk for jaundice, since they end up with more red blood cells to break down after birth.

But the data on jaundice are surprisingly mixed. Most studies have found no increase in bilirubin levels or need for phototherapy with delayed cord clamping.[35] However, the Cochrane review includes an unpublished dissertation (notably, by one of the review's authors) that found that babies with delayed cord clamping were more likely to need phototherapy to treat jaundice,[36] and these data play a major role in making this finding significant in the review.[37] Unfortunately, it isn't clear from the dissertation whether the attending pediatricians were "blinded to"(that is, were not given information on) the baby's cord clamping time, and actual bilirubin levels weren't reported. It's possible that this result was due to unintentional bias on the part of the pediatricians caring for the babies—expecting babies with delayed cord clamping to need phototherapy and therefore being quicker to prescribe it.[38] Because of this, some authors have questioned the conclusion that delayed cord clamping increases the risk of jaundice—showing that even a prestigious Cochrane review can be controversial.[39] This is an area that needs more research, but regardless, jaundice is always possible whether the cord was clamped early or late. All babies should be screened for jaundice around the third or fourth day of life and have access to phototherapy if needed.[40]

Some obstetricians also worry that delayed cord clamping might increase the mom's risk of postpartum hemorrhage, but studies show that whether the cord is clamped immediately or several minutes after birth doesn't affect maternal bleeding or the time it takes to pass the placenta.[41]

In most studies of delayed cord clamping, the baby was held at the level of the placenta for a two- to three-minute delay before clamping the cord, the thought being that transfusion from placenta to baby would be less effective if it had to work against gravity. Consequently, some obstetricians have hesitated to make delayed cord clamping their standard practice, because it gets in the way of letting the mom hold her baby skin-to-skin right away. As Dr. Tonse Raju, a neonatologist and chief of the Pregnancy and Perinatology Branch at the U.S. National Institute of Child Health and Human Development, commented in an editorial, "Trying to hold on to a wet, vigorously crying, and wriggling infant at the perineum for 2 minutes, in gloved hands, is awkward and can be risky.

When the mother is waiting anxiously to hold her baby and the father is taking photographs, two minutes can seem like an eternity."[42] However, a study published in the *Lancet* in 2014 put this fear to rest, showing that placental transfusion was no different whether the baby was held at the level of the perineum or on mom's chest or abdomen.[43] Your baby can go straight to skin-to-skin with you, and everyone can relax for a couple of minutes before cutting the cord.

OFFICIAL RECOMMENDATIONS AND THE PACE OF CHANGE IN OBSTETRIC PRACTICE

For premature babies, the benefits of delayed cord clamping are dramatic and clear, and delayed clamping is recommended by the American Congress of Obstetricians and Gynecologists (ACOG), the Society of Obstetricians and Gynaecologists of Canada, and the European Association of Perinatal Medicine.[44] Many hospitals are making at least a 30- to 60-second delay part of their standard protocol for premature babies, and pretty much everyone agrees that this is helpful.

The benefits of delayed cord clamping in full-term babies are subtler, and official recommendations vary. Since 2006, the World Health Organization (WHO) has recommended waiting at least one minute before clamping the cord for both full-term and preterm babies, and the International Liaison Committee on Resuscitation and the European Association of Perinatal Medicine agree.[45] However, in North America, both ACOG and its Canadian counterpart stop short of recommending delayed cord clamping for full-term babies.[46] The 2012 ACOG Committee Opinion states: "Currently, insufficient evidence exists to support or to refute the benefits from delayed cord clamping for full-term infants that are born in settings with rich resources."[47]

To understand ACOG's perspective, I talked to Dr. Jeffrey Ecker, who was vice chairperson of the ACOG Committee on Obstetric Practice when the Committee Opinion was written.[48] Reflecting on the inconclusive opinion, Dr. Ecker described the risks and benefits of delayed cord clamping in full-term infants as "the classic image of both sides of the scale." On one side is the benefit of improved iron stores. On the other side is the risk of jaundice. The ACOG committee saw these risks and benefits as being roughly in balance.

My interpretation is a bit different. The jaundice risk has only mixed data to support it, and I think the ACOG committee is misguided in downplaying the impact of iron deficiency in "settings with rich resources." While our hospitals may be rich with resources, the same can't be said for every mother and child. About 30% of U.S. women in their third trimester are iron-deficient, and the prevalence climbs still higher for black and Hispanic women.[49] Not every baby will be seen by a pediatrician in that critical 6- to 12-month window when iron stores may have dwindled. With 11% of 1-year-olds deficient in iron and rates much higher among some minority groups, iron deficiency is a significant public health issue. A study in Sweden, a country with a very low prevalence of anemia, found benefits of delayed cord clamping even in this advantaged population.[50] And waiting a couple of minutes to cut the cord is such a simple thing—effectively, a non-intervention. Why is there so much resistance to this change?

Dr. Ecker told me that ACOG would revisit its opinion on cord clamping every two years and that he thought hospitals would move in this direction. Meanwhile, he hoped that hospitals making the switch to delayed clamping would take the opportunity to study outcomes associated with the change. "I don't want to seem like a stuck-in-the-mud doctor," Dr. Ecker told me. "This is a doable thing. It happens in our delivery rooms and in delivery rooms across the country. I think it's going to become more and more common. I just think that we need to know more of the details."

CAMILA'S BIRTH STORY AND THE RESUSCITATION QUESTION

In 2012—by this time armed with a PhD in nutrition and her experience in the crowded Mexico City hospital—Camila Chaparro gave birth to her first child, a boy named Peter.[51] Throughout her pregnancy, and on the day of her delivery, Camila was cared for by a Certified Nurse Midwife practice in a university hospital in Washington, DC. The midwives were on board with everything that Camila and her husband wanted for their son's first moments in the world, all written down in their birth plan, including delayed cord clamping.

Despite their planning and preparation, Peter's first moments in the world were scary, and things didn't go according to plan. Peter was born with his umbilical cord wrapped tightly around his neck (called a nuchal cord), and he didn't breathe right away after birth. As Camila held her limp baby on her chest, she watched as the midwife quickly clamped and cut the cord. "And all of a sudden, there were, like, 15 people in the room, and they immediately transferred him over to the warmer," she told me. Nurses and neonatologists swarmed around her son, trying everything to help him breathe, and eventually needing to intubate him to get air to his lungs. "He didn't really start breathing on his own for the first 10 to 15 minutes, which was incredibly scary," Camila remembered.

Camila wonders, in retrospect, whether it might have helped Peter to leave the cord attached a bit longer. After all, it was pulsing with oxygenated blood. But at the time, she told me, this was the last thing on her mind. She just wanted to see her baby's chest rise and fall as air filled his lungs. (It did, and Peter is now a healthy, happy toddler.)

Camila's experience illustrates one of the most controversial points of the cord clamping debate and the area where we most need more research. It isn't just about iron or jaundice. Delayed cord clamping also throws a wrench into protocols for caring for newborn babies in trouble. And doctors love protocols—not because they're inflexible, but because protocols give everyone a job to do and a place to do it, and they help to prevent mistakes. Protocols might seem like overkill in a normal, uncomplicated delivery of a healthy newborn, but for a baby like Peter, the protocol is critical. And in most hospital delivery rooms, cutting the cord immediately is part of the protocol for helping a baby in trouble, so that he can be moved to a warmer to receive expert care by neonatologists and pediatricians. How does delayed cord clamping fit into that picture?

Some doctors and midwives argue that if a baby needs help breathing, that help should be given with the cord still attached.[52] In the case of a nuchal cord, like Peter had, by the time the baby is born, his blood volume and oxygen supply may already be very low. Waiting to cut the cord can help the baby recover some blood and oxygen, perhaps giving him a chance to breathe on his own or helping the resuscitation.[53] Neonatologist Tonse Raju described it to me like this: "People forget that the placenta actually continues to breathe for the baby even after the baby is born, as long as the umbilical cord is not severed. When the baby is in the mother's womb, it is the placenta that is doing gas exchange, and it will continue to do its job after the baby is born, even if the baby has not started breathing through his lungs . . . Therefore, in babies who require resuscitation, getting the extra blood may make the doctor's job of resuscitation much easier."[54]

The thing is, nobody really knows whether keeping the cord attached for a baby struggling to breathe is a help or a hindrance. Studies in lambs have found benefits of delayed clamping in such scenarios,[55] but it hasn't been studied in humans. It's no small burden to be a doctor caring for a baby who hasn't yet taken a breath, and resuscitating a baby with the cord still attached is an idea that probably puts lots of physicians out of their comfort zone, not to mention the logistical challenges of resuscitating a baby still tethered to his mom. To address this second problem, doctors in the United Kingdom have invented a rolling cart that gives a surface for resuscitation, as well as all the necessary equipment, while keeping the cord attached. Just as I finished this chapter, they published a paper describing its pilot use for 78 babies, most of them high-risk babies requiring resuscitation.[56] Most doctors who tried the trolley rated it the same as or better than standard resuscitation equipment, and most parents preferred having their babies right next to them during resuscitation. Another solution is to "milk" the cord several times to speed the transfusion of the blood to the baby, so that the cord can be cut relatively quickly. A few studies of cord milking have found that it offers benefits similar to delayed cord clamping, although more research is needed.[57] Regardless, changing delivery room protocols and figuring out where delayed cord clamping sits on the list of priorities among all the things that might happen at a birth is not a simple task, and it requires everyone in the delivery room—obstetricians, midwives, nurses, neonatologists, and pediatricians—to be on the same page.

CORD BLOOD AS A SOURCE OF STEM CELLS

Some researchers think that babies benefit from delayed cord clamping because cord blood is so rich in stem cells.[58] Because stem cells have been used to successfully treat many disorders and diseases, they argue, the stem cells in cord blood could help a newborn recover from the stress of birth and maybe even prevent diseases later in life—everything from asthma to autism to diabetes. This is an interesting hypothesis, but testing it would require very large studies that track children for many years. Those studies haven't been done, so at this point, this idea is purely speculative.

The potential utility of stem cells found in cord blood also means that you might think about collecting it for purposes other than the immediate use of the newborn baby. At the cost of thousands of dollars, you can store your baby's cord blood with a private stem cell bank, just in case your child or another family member might need a stem cell transplant in the future. However, the chance of privately banked cord blood being useful in the same child has been estimated as less than 1 in 2,700.[59] It's a waste of money for most families and is discouraged by the AAP, ACOG, the American Medical Association, and the American Society for Blood and Marrow Transplantation.[60] An exception might be if you already know that a stem cell–treatable disease runs in your family. In some cases, stem cells from a baby's cord blood might be useful to an older sibling already diagnosed with such a disease, such as some types of cancers and blood disorders.[61] If this is the case, there are sibling-donor cord blood programs that help guide parents through that process and make it affordable.[62]

Public stem cell banks accept and store donations of umbilical cord blood, and donating your baby's cord blood could help a future patient through a transplant or contribute to advancing stem cell research. If you choose this route, the cord may need to be clamped early in order to collect enough blood to be useful for a donation. This might be a reasonable choice for parents of a healthy full-term baby, so long as they pay close attention to iron nutrition later in infancy. However, some authors "question the ethics of using the newborn as a blood donor."[63] ACOG maintains that "the collection should not alter routine practice for the timing of umbilical cord clamping,"[64] the point being that cord blood donation or banking shouldn't compromise the health of the newborn "donor." Stem cell technology may soon allow for combining smaller samples from different babies or expanding single samples so that delayed cord clamping and cord blood donation don't have to be mutually exclusive. This is a rapidly evolving field, so check with your birth provider or a public cord blood bank to understand the current requirements if you're considering donating or banking cord blood.

WHERE ARE WE HEADED?

It will be interesting to watch how delivery room procedures change over the next few years. Timing of cord clamping might remain a gray area for a while—something parents can request and physicians will accommodate, with no clear recommendation. But there is no doubt that we'll continue to see more research on this fascinating question. This is part of the healthy push and pull of medical research and practice, and it is good that there is debate and critical assessment of the evidence. There may even be a shift in attitude or a change in ACOG's recommendation on cord clamping by the time this book is published. Regardless, real change in practice takes time, and I think we'll see variation in protocols on the timing of cord clamping for a while. If you're interested in delayed cord clamping, it is worth discussing with your obstetrician or midwife ahead of time, including consideration for situations that might require immediate clamping.

3 ⁂ OF INJECTIONS AND EYE GOOP

Newborn Medical Procedures

When I think back to the moments after Cee's birth three years ago, there are a few things that stick in my mind very clearly: the quiet on her face as we gazed at each other, her confidence in her first breastfeed, and the exhilaration of being on the other side of childbirth. But a lot of other things were going on during that time, and they're mostly a blur. There was the placenta to deliver and a few stitches for me, and for Cee, a bunch of routine medical procedures that happened in the first hours and days of her life—a quick succession of heel sticks, screening tests, injections, and eye goop. These procedures can feel a little intrusive during this time when your baby is adapting to the outside world and you're trying to get to know one another. As parents, we want to understand the risks and benefits of these procedures and whether they are truly necessary. In this chapter, I explore the science of two recommended newborn procedures in depth: the vitamin K shot and erythromycin eye ointment. Several others are summarized at the end of the chapter. Know that with all of these newborn procedures, you can ask that they be delayed for at least the first hour after birth so that your getting-to-know-you time is uninterrupted.

VITAMIN K INJECTIONS

Why Vitamin K Is Important: One Baby's Story

Olive Eloise Leavitt slid into the world early one morning in January 2014 at a birthing center in Washington state. Her mom, Stefani, had labored without medication through the night. When it was time to push, Stefani climbed into the birthing tub, pushed through three contractions, and then had her healthy little girl in her arms. As for most newborns, Olive's first month was a sweet blur of sleeping and breastfeeding, while her parents and big sister got used to being a family of four.[1]

But a month later, on Valentine's Day, Stefani started to worry about Olive. They'd been up for most of the night, struggling to breastfeed, and now Olive seemed sleepy and completely uninterested in eating. After an afternoon nap, she was so lethargic that she was barely able to open her eyes. Stefani took her to the emergency room.[2]

What happened next is a parent's worst nightmare. What at first seemed like a minor concern was suddenly a life-threatening emergency. A lumbar puncture (spinal tap) showed blood in Olive's spinal fluid, and she was rushed to a better-equipped hospital. It took many attempts to get an IV placed, because Olive's fragile veins kept rupturing. Finally, a CT scan revealed the biggest problem of all: a huge mass of blood filled Olive's skull and pushed her brain to the side. The doctors told Stefani and her husband that Olive had an intracranial hemorrhage, and if she survived, there was a good chance that her brain would be severely damaged.

At some point, as the doctors worked on Olive, one of them asked Stefani if her baby had received a vitamin K shot at birth. The answer was no, and this explained everything. Like all newborn babies, Olive was born with very little vitamin K in her body. Vitamin K is essential for the formation of blood clots, and Olive didn't have enough of it. This was why it was hard to establish an IV, and it was why she had an intracranial hemorrhage. A simple injection of vitamin K at birth would have prevented Olive's pain and suffering, the fear and anxiety of her parents, and the medical expense of her treatment.

Olive needed brain surgery, but first, she needed to be able to clot her blood; otherwise, the surgery would cause even more bleeding. She received an IV infusion of vitamin K, and after about an hour, her blood began to clot. Doctors worked through the night to remove the mass of blood from her brain. In the days that followed, Olive's parents sat by her side and watched, not knowing how much damage might have been done to her brain. They watched as she moved her leg for the first time, as she opened her eyes, and finally, after her breathing tube was removed four days later, as she cried. "I have never been happier to hear a baby cry," Stefani wrote on her blog.[3]

Two weeks later, Olive was discharged from the hospital. Her paperwork included a long list of diagnoses: "Hemorrhagic Disease of the Newborn Due to Vitamin K Deficiency, Subdural Hematoma, Acute Respiratory Failure, Increased Intracranial Pressure, Ischemic Brain Damage, Seizure, Cerebral Infarction of the Left Hemisphere, and Cholestatic Jaundice." As I write this, it has been three months since Olive's surgery. Her parents and doctors continue to watch her closely, but they haven't yet seen any signs of lasting brain injury.[4] She is incredibly lucky.

That Olive didn't get her vitamin K shot seems to have been an unfortunate oversight at the birthing center. Some parents, however, intentionally decline the vitamin K shot for their newborns. They worry about the pain of the injection and that it's an unnatural and unnecessary intervention. Some may have heard that the vitamin K shot can cause cancer.[5] There is a ton of misinformation about vitamin K on the Internet to fuel these concerns, and at the very least, it leaves caring parents confused. When I dug into the science on this topic, I found a fascinating history, intriguing physiology, and a clearer understanding of the benefits and risks of the vitamin K shot.

The History and Science of Vitamin K

Boston doctor Charles W. Townsend published a description of 50 cases of strange bleeding in newborns, many of them fatal, in 1894.[6] The cause was unknown, and all that doctors could do was try to control the bleeding, hoping that the blood would eventually start to clot. At that time, nobody knew what vitamin K was, much less that it could stop these babies from bleeding.

The discovery of vitamin K was an accident. In the 1930s, a Danish biochemist named Henrik Dam was feeding baby chicks low-fat diets as part of his work on cholesterol metabolism. He noticed that some of the chicks were bleeding in their muscle and under their skin.[7] In a series of careful experiments, Dam substituted different types of foods in the chicks' diet. He found that the bleeding occurred when he fed them only foods such as rice, sunflower seeds, corn, and rye, but it was completely prevented by kale, hemp seed, and pig liver. Dam eventually isolated the protective factor from alfalfa leaves, and he and other researchers showed that it could stop hemorrhaging not just in baby chicks but also in newborn infants. Dam named this fat-soluble compound vitamin K, for "Koagulations."[8] For his work on vitamin K, Dam shared the 1943 Nobel Prize in Physiology or Medicine with Edward Doisy, an American biochemist who studied the chemistry of the vitamin.[9]

The ability of blood to clot is essential to mammalian life. Our bodies are full of blood, pumping through the vast network of the circulatory system, but this system works only if the blood stays within the walls of the blood vessels. If one of these vessels is damaged—not hard to do, since their walls are thin and fragile—immediate clotting of the blood is what prevents hemorrhage.

Vitamin K is required for the liver's synthesis of four proteins that contribute to the ability of blood to clot.[10] Some of the richest sources of vitamin K_1 (also called phylloquinone) in the human diet are foods such as leafy green vegetables (kale, Swiss chard, collard greens), cruciferous vegetables (Brussels sprouts, cabbage, broccoli), and soybean and canola oils. Vitamin K_2, a group of similar compounds (called menaquinones), is made by bacteria, including gut microbes in humans. It isn't clear what role menaquinones play in human nutrition, because they are not well absorbed from the gut.[11]

When it comes to vitamin K, babies are born at a disadvantage. During pregnancy, very little vitamin K crosses the placenta from mom to baby; cord blood levels of the vitamin are so low that they're often undetectable.[12] Even if a mother takes large amounts of oral or injected vitamin K before giving birth, her baby is still born with very little vitamin K.[13] After birth, the vitamin K supply doesn't improve much. Breast milk is very low in vitamin K,[14] and the nascent bacterial population in a newborn's gut doesn't contribute much in the way of vitamin K_2.[15] Although this will increase in the first weeks of life, the bacteria that produce the most vitamin K, *E. coli* and *Streptococcus*, are more prevalent in formula-fed infants than breastfed infants.[16]

Most newborn babies do okay despite these low vitamin K levels, but some develop one of three types of vitamin K deficiency bleeding (VKDB).[17]

- Early VKDB occurs within the first 24 hours after birth. It is rare, but when it does occur, it's almost always in babies whose moms were taking medications like blood thinners or anticonvulsants during pregnancy.
- Classical VKDB occurs between 2 and 7 days of age. Bleeding is

often seen at the umbilical cord, from the skin or nose, from the gastrointestinal tract, or from the site of circumcision. Estimates suggest that, without receiving vitamin K at birth, 1 to 7 babies in 400 will have classical VKDB. So long as medical care is available, most babies will fully recover.

* Late VKDB, the type that affected Olive, occurs between 8 days and 6 months of age. Although rare (4 to 7 in 100,000 babies), late VKDB can be devastating. In one analysis of published cases, 63% had severe brain hemorrhage and 14% died. Among those that survived, 40% had lasting neurological damage.[18]

Classical and late VKDB are completely prevented by a dose of 1 mg of vitamin K given as an intramuscular injection at birth, as recommended by the AAP since 1961.[19]

Most cases of classical and late VKDB occur in exclusively breastfed babies. Breast milk has just 1 to 2 μg per liter of vitamin K, lower than

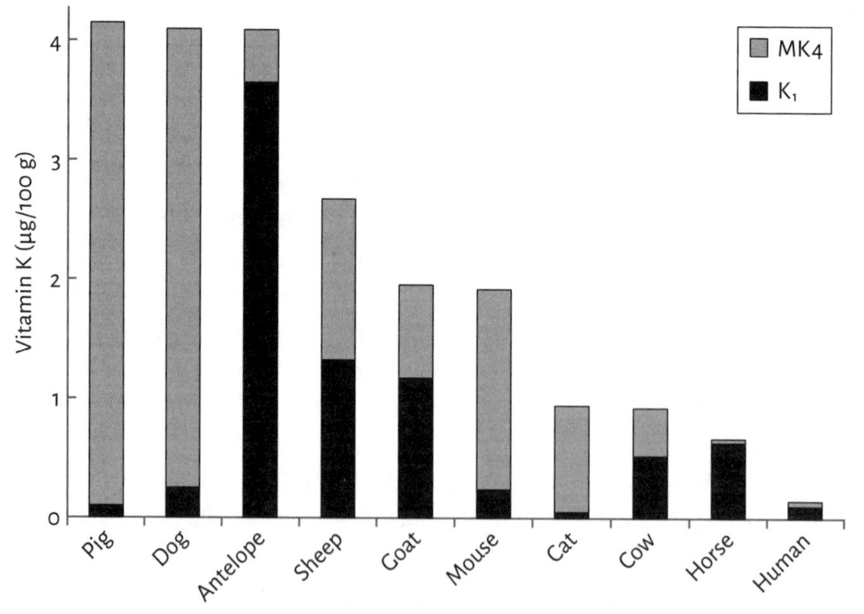

Vitamin K concentrations in milk from various mammals. The two major types of vitamin K found in milk are phylloquinone (K_1) and menaquinone-4 (MK4, a type of K_2). *Source:* H. E. Indyk and D. C. Woollard, "Vitamin K in Milk and Infant Formulas: Determination and Distribution of Phylloquinone and Menaquinone-4," *Analyst* 122 (1997): 465-69.

in other animal species. Infant formula contains at least 55 μg per liter, enough to prevent VKDB in infants that are at least partially formula-fed.[20] (A microgram, μg, is one-thousandth of a milligram.)

Without the vitamin K shot, breastfed newborns, especially those that have early feeding difficulties (or if the mom's milk is slow to come in), can rapidly deplete their small vitamin K reserves, leaving them susceptible to classical VKDB.[21] Late VKDB almost always occurs in apparently healthy, exclusively breastfed babies.[22] Note that the incidence of VKDB given above for each type of VKDB reflects the data for entire populations of babies, many of whom received some formula. Since late VKDB usually only affects exclusively breastfed babies, the true incidence in these babies is probably much higher.

Babies born with liver diseases that impair the secretion of bile, necessary for the absorption of fat-soluble vitamins, are at especially high risk for late VKDB. This turned out to be the case with Olive, although her liver disease wasn't discovered until more than a month after her brain hemorrhage.[23] Unfortunately, there often aren't any signs of the disease before VKDB happens. Vitamin K at birth spares these most vulnerable babies from the severe effects of late VKDB.[24]

There is some evidence that we're seeing more cases of VKDB because parents are declining the vitamin K shot, and perhaps in part because more babies are exclusively breastfed. During just eight months in 2013, pediatricians at Vanderbilt University Medical Center were alarmed to see five babies with late VKDB, four with brain hemorrhages.[25] There hadn't been a case of late VKDB at Vanderbilt in at least five years. In all five cases, these were healthy breastfed infants who had not received the vitamin K shot at birth. In 2013, 3% of babies born at the Vanderbilt hospital did not receive vitamin K at birth, but 28% of those born at local birthing centers missed the shot.[26] Mothers who give birth at birthing centers are often more interested in minimizing interventions around childbirth and in breastfeeding, and it is a cruel irony that the same babies who may not receive vitamin K at birth are also the most vulnerable to VKDB.

The Vitamin K and Cancer Scare

One of the reasons parents give for declining the vitamin K shot is that they've heard it might cause cancer, and there's an interesting history behind this concern. In 1992, British epidemiologist Jean Golding and her colleagues published a paper in the *British Medical Journal* reporting an association between childhood cancer and the vitamin K shot, but not oral vitamin K (commonly used in the United Kingdom at the time).[27] In response, the journal was flooded with letters to the editor from doctors and researchers, most criticizing the study.[28] They pointed out flaws in the study design, including issues of bias and possible inaccuracies of the data. Researchers from the United Kingdom and the United States noted that cancer rates had not increased over time, even as vitamin K administration had become more common.[29] These limitations put the study's conclusions on shaky ground, but the media reports focused on the scary cancer connection, not the problems with the science. Understandably, parents were afraid.

Golding's study showed a correlation between two variables, but it could not prove that vitamin K caused cancer. And so, more studies were done. At least 12 more studies were conducted to look at this question, and together they found that a link between cancer and vitamin K is extremely unlikely.[30] Meanwhile, there is a certain outcome for not giving vitamin K: babies can die or become severely disabled because of VKDB.

This story illustrates an important point about scientific evidence that I mentioned in chapter 1: one study is rarely enough to be conclusive. What's important is the consensus of multiple studies, conducted by different researchers, in different places, using different methodology. Golding's study has since been overshadowed by many more studies finding no link between cancer and vitamin K. However, we can still find scary articles on the Internet claiming that the vitamin K shot causes cancer, complete with peer-reviewed research from a well-respected journal. To be savvy consumers of information, we have to look at all of the research together. There is a strong scientific consensus that vitamin K doesn't cause cancer. Anyone who claims otherwise is cherry-picking from the evidence, choosing one study that supports the claim and completely ignoring the rest.

The Scoop on Oral Vitamin K

Everyone agrees that a single intramuscular vitamin K shot at birth prevents both classical and late VKDB. However, because of the cancer scare and because nobody wants to poke a newborn baby unless it is necessary, many countries have tried various methods of oral dosing.

Since VKDB is so rare, it isn't practical to conduct randomized controlled trials to determine the most effective dosing protocols. Instead, countries that use oral dosing have just tried different plans—like one dose at birth or multiple doses during the first months of life—and then tracked the number of VKDB cases before and after a change in plan. This gives us surveillance data, and when we compare the results from different countries using different plans, we can see what worked and what didn't.

Together, these data clearly show that intramuscular vitamin K is most effective at preventing VKDB. Oral dosing plans, on the other hand, have had mixed success, often preventing classical but not late VKDB. However, a few oral dosing schemes do seem to work well. In Denmark, giving 2 mg orally at birth, followed by a weekly 1 mg dose as long as breastfeeding continues, prevented late VKDB, even in babies with liver disease.[31] A recent report from Switzerland found that VKDB was prevented by a 2 mg dose at 4 hours, 4 days, and 4 weeks after birth (a protocol chosen in part because it is easy to remember).[32] During the six years of the Switzerland study, there were four cases of late VKDB, but in all of these, the parents had either refused vitamin K or forgotten to complete the doses. These results are promising, but since they come from just a few years of births in relatively small countries, it will be important to continue tracking VKDB with these dosing plans.

Oral doses are tricky, for a couple of reasons. First, unlike other fat-soluble vitamins, vitamin K doesn't stick around in the body very long. In adults, the body's vitamin K_1 supply turns over in a matter of a day or two.[33] This explains why multiple oral doses are necessary. Second, absorption of an oral dose from the digestive tract into the body is highly variable even among healthy babies and especially among babies with liver disease.[34] On the other hand, when vitamin K is injected, it seems to be stored in the muscle and slowly absorbed into the blood over time, providing a continuous supply for the first few months of life.[35]

Despite the success of Denmark's and Switzerland's oral dosing plans, we will probably continue to see the intramuscular injection recommended in many parts of the world, including the United States. Why? Because it works, and it is safe. Plus, oral dosing relies on parents remembering to give it weeks and months later, and new parents have been known to forget a few things in the haze of early parenting. Oral dosing is also more expensive, requiring at least six times the amount of vitamin K per baby.

Some hospitals in the United States, including Stanford's Lucille Packard Children's Hospital,[36] offer to give vitamin K orally if parents refuse to allow their newborns to receive the shot. This means using the injectable vitamin K preparation off-label, since there currently isn't a vitamin K product licensed by the U.S. Food and Drug Administration for oral use in babies. Products marketed for babies are available to purchase online, but it's important to remember that the supplement industry in the United States is poorly regulated. Unlike the process for FDA approval for drugs, supplement manufacturers don't have to prove that their products are safe, effective, and without contaminants, or even that they contain the amount of vitamin K they report on the label.[37]

Among the misinformation about vitamin K that you can find online is the suggestion that eating lots of kale or taking vitamin K supplements during pregnancy and while breastfeeding can prevent VKDB. Remember that vitamin K barely crosses the placenta in pregnancy, even if you take large amounts of supplements.[38] As for breast milk, eating a diet rich in food sources of vitamin K or taking a normal supplement results in little to no increase in the vitamin in breast milk.[39] However, in one study, mothers took a very large vitamin K supplement (5 mg/day), which raised breast milk concentrations to a level similar to that in infant formulas, which we know prevents VKDB.[40] However, their infants still had plasma vitamin K levels lower than those of formula-fed infants. Five milligrams is also a lot of vitamin K. Standard vitamin K supplements sold in my grocery store contain just 0.1 mg, and obtaining 5 mg per day through diet would require you to eat more than 10 cups of chopped kale daily.[41] Another problem is that this method might fail if either mom or baby has low vitamin K absorption. In other words, trying to prevent VKDB through your diet might be possible, but it's a roll of the dice compared with the certainty of the vitamin K shot.

What Ingredients Are Found in the Vitamin K Shot? Are They Safe?

As in any medication and many foods, the ingredients in the newborn vitamin K shot can seem intimidating at first glance. In addition to vitamin K, the shot also includes several "inactive" ingredients. These have

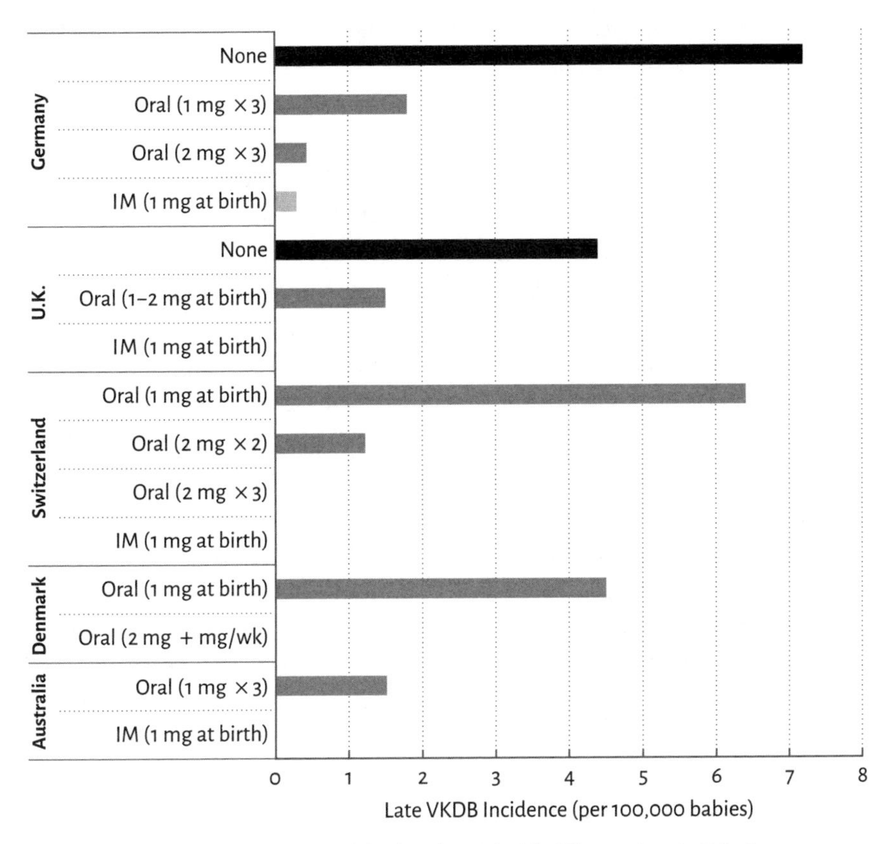

Incidence of late vitamin K deficiency bleeding (VKDB) with different vitamin K dosing protocols. Values do not include cases where parents refused vitamin K or did not complete the dosing series. An intramuscular (IM) injection of 1 mg of vitamin K at birth offers complete or near-complete protection from VKDB. Protection by oral doses varies. For oral dosing protocols, "× 2" indicates that doses were given at birth and again in the first week, and "× 3," that a third dose was given after about a month. *Sources:* M. Cornelissen et al., "Prevention of Vitamin K Deficiency Bleeding: Efficacy of Different Multiple Oral Dose Schedules of Vitamin K," *European Journal of Pediatrics* 156 (1997): 126; B. Laubscher et al., "Prevention of Vitamin K Deficiency Bleeding with Three Oral Mixed Micellar Phylloquinone Doses: Results of a 6-Year (2005-2011) Surveillance in Switzerland," *European Journal of Pediatrics* 172 (2013): 357-60; M. J. Shearer, "Vitamin K Deficiency Bleeding (VKDB) in Early Infancy," *Blood Reviews* 23 (2009): 49-59.

unfamiliar chemical names, and you might feel worried about injecting them into your newborn.

To make matters worse, if you read the package insert for the vitamin K shot, you'll see lots of dire warnings about the ingredients. This makes great fodder for scary Internet articles, which is exactly what you'll find if you Google "ingredients in vitamin K shot." However, the package insert is required to include warnings about *any* cases of toxicity linked to ingredients in the shot, even if they occurred in different drugs, in different doses, and in very different types of patients. For example, most of the warnings listed on the vitamin K label refer to critically ill patients or premature babies receiving much larger doses of these ingredients intravenously and over many days. The package insert ensures that you're fully informed of every possible risk, but most of it is irrelevant to you and your baby. (The same is true of vaccine package inserts.)

Knowing that the package insert wasn't of much use, I started digging through toxicological research to find more information about the ingredients. I found that every ingredient is there for a reason. Some keep the vitamin K in the injectable solution. Vitamin K is fat-soluble, so just as oil and water don't mix, it can't simply be dissolved in water or saline for the injection. Other ingredients are used to keep the pH of the solution in a neutral range or to prevent bacterial growth, important aspects of safety of the shot. Any chemical can be toxic in a high enough dose, which is why the package insert for vitamin K is a scary read. But when I looked at the amounts of each ingredient used, it was clear that there is no reason to fear that the shot—given as a one-time intramuscular dose—is toxic to newborns. For those who want to learn more, I have included a detailed description of each ingredient and relevant toxicological data in appendix A.

There are just a few possible side effects of the vitamin K shot. The most immediate is the pain of the injection. However, this is brief, and you can help ease it by holding your newborn skin-to-skin and breastfeeding during the shot.[42] Among the millions of doses of vitamin K given to newborns, only one case of anaphylactic shock due to an intramuscular vitamin K injection has been reported. This baby was treated and fully recovered.[43]

Is There an Evolutionary Explanation for Vitamin K Deficiency?

It's tempting to look at the relative vitamin K deficiency in human babies and wonder whether there is an evolutionary explanation for the poor placental transport, low breast milk concentration, and limited production by gut bacteria in breastfed babies. Could there have been an evolutionary advantage to low vitamin K in our ancestors, even if it also increased their risk of hemorrhage?

"There probably was, but whatever it was, it doesn't exist anymore." That's what Dr. Frank Greer, a neonatologist and professor of pediatrics at the University of Wisconsin School of Medicine and Public Health told me when I posed this question to him.[44] He's been doing research on infant nutrition, including vitamin K, for over 30 years, and he hasn't figured out this mystery. He wonders whether low vitamin K might have protected babies with severe birth trauma and asphyxiation, which today would probably be delivered by C-section, from developing disordered blood clotting. "But I'm just wildly speculating," he said. "Whatever it is isn't an advantage anymore, because nobody has observed *any* harm from the vitamin K shot."

Other researchers speculate that perhaps ancestral babies did have a bit more vitamin K, maybe because their moms ate more leafy greens and had more vitamin K–producing gut bacteria.[45] Although it's unlikely that these factors could completely prevent VKDB, they may have reduced the incidence. It's also important to remember that VKDB is rare, and among all the other dangers facing early human babies, it probably couldn't exert much evolutionary pressure. It's interesting to think about evolution and vitamin K, but it doesn't change what modern science has clearly revealed: giving your baby a shot of vitamin K at birth effectively and safely prevents a rare but potentially devastating bleeding disorder.

NEWBORN EYE OINTMENT

Why the Eye Ointment?

Whenever one of my friends has a new baby, my Facebook feed is usually graced with a photo of the red-faced little one. Even though each of these babies brings to the world unique genetics, facial features, and personality, their photos look remarkably similar. They're almost always wrapped in a gender-neutral pink and blue flannel blanket, and their eyes are shiny with a thin layer of ointment. What is up with that eye goop?

If we want to understand why we put ointment in the eyes of newborn babies today, we have to begin our story in the maternity hospitals of nineteenth-century Europe. One of the major threats to newborn health during this time was an eye infection, called neonatal ophthalmia, that developed soon after birth. It was usually caused by gonorrhea, a sexually transmitted disease carried by 35% of women giving birth in hospitals at the time and passed from mom to baby during childbirth. Newborn eye infections were such a problem that hospitals had separate wards for babies showing signs of infection. If a woman had gonorrhea, her baby was placed in a crib that was out of reach of the mother's bed, and it was only when a nurse brought the baby to mom for feeding that the two were allowed to be together.[46] This was before the discovery of antibiotics, so there wasn't an effective treatment for these infections. Gonococcal eye infections (caused by gonorrhea) left nearly a quarter of infected babies with lasting vision damage. It was the biggest cause of blindness in infancy at the time; half of the children in European schools for the blind were there because of gonorrhea.[47]

Dr. Carl Credé, a German obstetrician and director of the maternity hospital in Leipzig, was determined to improve this situation. In his hospital, nearly 14% of newborn babies developed neonatal ophthalmia. Recognizing that babies were being infected during labor, Credé's first idea was to clean the vaginas of infected women really well. He went so far as to apply douches of carbolic or salicylic acid every 30 minutes while women were in labor. This seemed to prevent some infections, but the incidence still hovered around 10%, and Credé wasn't satisfied.[48]

Next, Credé tried treating the babies' eyes at birth. He put a single drop of silver nitrate solution in each eye, and he found that it almost completely prevented neonatal infections in babies whose moms had gonorrhea. Knowing that many women had asymptomatic infections, he began applying silver nitrate drops to the eyes of every baby born in his hospital, and thankfully, also stopped douching women during labor. In 1883, he reported that of 1,160 babies treated with silver nitrate under his care, there were only 2 cases of neonatal ophthalmia. He urged other obstetricians and midwives to try this prophylaxis (a medical treatment meant to prevent disease), and it was adopted around the world, preventing blindness in countless babies.[49]

Newborn Eye Prophylaxis in the Modern Era

In many countries around the world, including the United States, newborns are still given eye prophylaxis soon after birth, but the circumstances are different from in Credé's time. Most importantly, the incidence of gonorrhea has dropped.[50] Chlamydia and other types of bacterial infections now cause most cases of neonatal ophthalmia. These cases are not nearly as severe as those caused by gonorrhea, and they don't generally cause blindness.[51] We also now have antibiotics, so unlike in Credé's day, these infections can be treated. Nobody ever wants to see a new baby get an eye infection, but the outcomes aren't nearly as scary as they used to be.

Another thing that has changed is that we have excellent tests to screen for gonorrhea and chlamydia. Ideally, women that are at risk for these diseases are screened during pregnancy, and if positive, they and their sexual partners are treated before the baby is born. This approach improves the health of the woman and her partner *and* minimizes the risk to the baby.[52]

Finally, the type of eye treatment used has changed. Credé's silver nitrate was eventually abandoned because it caused chemical conjunctivitis—inflammation of the eyes—in 50% to 60% of babies. This went away on its own within a couple of days, but still, it was probably uncomfortable for the babies and distressing for parents.[53] In the United States and Canada, among other countries, erythromycin (an antibiotic) ointment is now used for eye prophylaxis. Erythromycin rarely causes chemical conjunctivitis. It is effective against gonorrhea, and some studies suggest that it also prevents chlamydial infections, although the data are mixed on this point. A downside to erythromycin is that, because it is an antibiotic, its frequent use may encourage the development of erythromycin-resistant bacteria.[54] (More on this later.)

Povidone-iodine, an antiseptic, is also used for eye prophylaxis in many parts of the world. It is effective against both gonorrhea and chlamydia, as well as herpes simplex virus.[55] It causes little to no chemical conjunctivitis and doesn't cause antibiotic resistance. It's also inexpensive. Many consider it the best option for eye prophylaxis. However, not all studies are so positive, and more research is needed.[56] Although povidone-iodine has been approved for use in the United States by the FDA, no commercial product is yet available.[57]

Eye prophylaxis was a slam-dunk for public health in the time of Credé. Today, it isn't so clear. The problem is not as grave, and with antibiotics and good screening tests, we have other ways to prevent and treat these infections. Does it still make sense to treat every baby with eye prophylaxis at birth? In the next few sections, we'll look at the benefits and risks of this treatment.

The Case for Eye Prophylaxis

The WHO, CDC, and AAP all recommend newborn eye prophylaxis, primarily for the purpose of preventing eye infections caused by gonorrhea.[58] The incidence of gonorrhea has decreased in recent years, but it remains a significant public health problem. In 2012, about 172,000 cases of gonorrhea were identified among women in the United States, but the real incidence is probably twice this since many cases go undiagnosed.[59] Most women with gonorrhea have no symptoms and don't realize that they are infected.[60]

The best way to prevent neonatal eye infections is to diagnose and treat women with gonorrhea and chlamydia during pregnancy. However, there are a few barriers to that goal. It would be too expensive to test every single woman in pregnancy, so the CDC recommends that doctors test only women that have risk factors for sexually transmitted diseases. And although the screening tests are very good, they are not 100% accurate, so some infections might be missed.[61] It's also possible for a woman who tested negative for gonorrhea early in pregnancy to be infected later in pregnancy (for example, if she or her partner is having sex with someone else), and that's not the kind of information that is usually shared at the birth of a baby. And finally, many moms don't have any prenatal care. All of these factors leave some babies vulnerable to gonococcal infection at birth.[62]

If a woman has gonorrhea at the time of childbirth, her baby has a 30% to 40% chance of developing an eye infection caused by the disease.[63] This usually occurs during a vaginal birth, but it can also happen to a baby born by C-section if the fetal membranes were broken before the surgery.[64] When looking at all causes of eye infections (not just gonorrhea), several studies have shown that babies born vaginally or by cesarean get the same types of infections at the same rates, and eye prophylaxis reduces infections regardless of mode of delivery.[65]

Although rare, a gonorrhea infection can be very serious to a baby, and it isn't a fair way to start life. Treatment requires hospitalization and injected antibiotics.[66] A gonococcal eye infection can not only cause permanent vision damage but also serve as an entry point for a systemic infection, which can cause septicemia, arthritis, and meningitis.[67]

Universal eye prophylaxis is a public health strategy in the United States. It is understood that most babies aren't actually at risk for gonococcal eye infections. "We are treating a lot of babies so that we don't miss one or two who might otherwise get a devastating disease," Dr. Kristi Watterberg, chairperson of the AAP's Committee on Fetus and Newborn, told me.[68]

Some countries, including Denmark, Sweden, and the United Kingdom, have stopped universal eye prophylaxis in newborns, opting instead to count on prenatal screening and treating only those babies judged to be at higher risk. However, this change in strategy resulted in an increase in the number of newborn gonococcal infections in these countries.[69]

Sometimes, when Americans hear that European countries are approaching public health in a different way, we like to think that the European way must be better. But national public health strategies aren't one-size-fits-all; they're customized to the disease prevalence and unique challenges within a population. For example, in 2010, Denmark, Sweden, and the United Kingdom reported 4.3, 4.7, and 18 cases of gonorrhea, respectively, per 100,000 women.[70] In the United States there were 108 cases per 100,000 women that year.[71] The United States has a larger and more diverse population with more disease, and we want to protect newborns from infection with that disease.

The case for eye prophylaxis rests on protecting babies that are often born into the least fortunate situations and with limited access to medical care. Erythromycin ointment is inexpensive, and applying it to babies' eyes is quick and easy. But if we're going to justify a prophylactic treatment, we also need to be sure that it is very low risk.

What Are the Risks?

The risks of erythromycin eye ointment are minimal. It occasionally causes chemical conjunctivitis, but this resolves on its own within a day or two.[72] There isn't much information on the actual prevalence of chemical conjunctivitis caused by erythromycin, but Dr. Margaret Hammerschlag, director of Pediatric Infectious Diseases at SUNY Downstate Medical Center, told me that it is so uncommon that "it's a non-issue."[73] A local neonatologist told me that he couldn't recall ever seeing a case of it in more than 10 years of practice.[74]

One concern about eye prophylaxis is that it might affect a new baby's vision, getting in the way of bonding with mom and dad after birth.

Studies done in the late 1970s and early 1980s in Colorado observed the behavior of babies and their parents with and without silver nitrate eye drops.[75] Compared with babies who got silver nitrate soon after birth, those who hadn't yet received it had their eyes more wide open and were more likely to visually follow a shape that was passed across their field of vision. What was more interesting, however, was the behavior of the parents. When their baby's eyes were wide open (without silver nitrate), the moms smiled at the baby more, and the dads spent more time looking at their babies, affectionately touching them, picking them up, and talking to them. Eye prophylaxis might not matter too much to the baby, because babies can't see that well at birth, and they're using their other senses to get to know us (see chapter 4). But whether or not a baby's eyes are open does seem to matter a lot to parents. Eyes are a meaningful way of making a connection with another person, including a newborn baby, and wide-open eyes seem to invite more of a connection.

Unfortunately, these studies were done using silver nitrate, and there isn't any research to tell us whether these same effects are seen in newborns treated with erythromycin. Regardless, it might be wise to delay your baby's eye ointment until after your first hour and first feeding together. By this time, your baby may be getting sleepy and will have had enough interaction for a while anyway. The AAP supports up to a one-hour delay in eye prophylaxis.[76]

There are a couple of possible big-picture risks of newborn eye prophylaxis as well. One is that the widespread use of any antibiotic could encourage the development of antibiotic-resistant microbes. However, after several decades of the use of erythromycin for this purpose, antibiotic resistance doesn't seem to be a problem. I was able to find only one published report of an outbreak of erythromycin-resistant eye infections (caused by *Staphylococcus aureus*) in a newborn nursery, occurring in Minnesota in the late 1980s.[77]

Another possible concern with the prophylactic use of erythromycin eye ointment in newborns is that it might affect the development of their microbiomes—the large, diverse community of mostly beneficial microbes that live in and on our bodies. In a newborn baby, the skin and gastrointestinal tract are rapidly colonized with microbes, and antibiotic use at birth or soon after birth has been shown to alter the developing microbiome.[78] However, this has been shown only in babies that received multiple doses of oral or IV antibiotics. Whether a very small amount of antibiotic ointment applied to a baby's eyes just one time could have an impact on microbes in the gut would depend on whether that antibiotic is absorbed systemically—into the blood—in a sufficient quantity. Is this plausible?

There is a bit of evidence to suggest that this is indeed plausible. First, we know that eye medications can be absorbed systemically, though this has never been studied specifically for erythromycin ointment.[79] A Texas study found that babies that received erythromycin eye ointment had their first poopy diaper more than two hours earlier than those that received silver nitrate.[80] As it happens, other studies have shown that erythromycin increases intestinal motility, or contractions of the intestine, including in infants.[81] The finding that erythromycin ointment appeared to speed up the babies' first bowel movement supports the idea that enough of the antibiotic might be absorbed to affect the gut. However, this was an observational study—not randomized—and it has never been replicated.

The idea that erythromycin eye ointment might affect the microbiome is just a hypothesis, and so far, we have little evidence to support it. However, there is interest in this question among microbiome experts. Dr. Henry Redel, in collaboration with Dr. Martin Blaser, director of the Human Microbiome Program, is planning a study to investigate it.[82] Perhaps they'll learn that erythromycin does affect an infant's microbiome—maybe just a little, maybe just temporarily. Or maybe they'll find no effect at all. The neonatal microbiome is a hot area for research right now, and among the intriguing research emerging there are undoubtedly some dud hypotheses. I'm skeptical about this one, but I will be watching for more data to be published.

But What About *My* Baby?

We've talked about how newborn eye prophylaxis is helpful from a public health perspective, but you're probably more concerned with your own very special baby. If you know you don't have gonorrhea (because you were tested in pregnancy and you've been in a trusted monogamous long-term relationship), then your baby has a very low risk for catching a nasty eye infection during childbirth. Does it make sense for your baby to get erythromycin ointment?

I think you could argue this either way, and how you feel about it probably comes down to how you think about risk and your comfort level with medications. On the one hand, erythromycin ointment is most likely harmless to your baby, and it just might prevent an eye infection. On the other hand, if you know your baby has a low risk for infection, it would also be reasonable to opt out of routine eye ointment and treat your baby only if an infection does arise.

There's just one problem with this choice as I've presented it. In many states in the United States and provinces in Canada, it isn't actually a choice. Eye prophylaxis is sometimes mandated by law, and it can be difficult to opt out.[83] In my opinion, this is too bad, if only because it makes parents feel bullied and frustrated that the decision isn't in their hands. If you're interested in declining the erythromycin, you'll want to talk with your obstetrician, pediatrician, and the birthing facility before giving birth so that you know what to expect. (In our local hospital, opting out just means a brief discussion with a nurse and signing a form, but local policies vary quite a bit.)

OTHER COMMON NEWBORN PROCEDURES

Other newborn medical procedures include several screening tests that are recommended for all babies within the first few days after birth:[84]

* *Screening for inherited diseases:* Between 24 and 48 hours after birth, your newborn's heel will be pricked with a needle to collect a few drops of blood. From this small sample, up to 50 inherited or congenital disorders can be detected (the exact number depending on state requirements), including metabolic, hormonal, blood (such as sickle-cell anemia), immune, and enzyme disorders. These are serious diseases that can cause cognitive impairment, slowed growth, seizures, and even death, but early treatment allows children to lead longer and healthier lives.
* *Hearing test:* One to two newborns per 1,000 are born with hearing loss. Hearing is an important way for babies to learn, and hearing loss can get in the way of normal language and cognitive development. With early intervention, such as speech therapy, sign language, and/or cochlear implants, babies can develop normal language skills. The hearing test takes about 5 to 15 minutes, is noninvasive, and can even be completed while your baby is asleep.

⁎ *Congenital heart disease screening:* Six to 13 newborns per 1,000 are born with congenital heart disease, and the sooner it is identified, the better the prognosis for treatment. This test is very simple: a painless sensor is placed on the baby's hand or foot to measure heart

⁎ RECOMMENDED INFANT HEALTH RESOURCES

Beginning during pregnancy, new parents-to-be have lots of questions about infant health, from the common medical interventions and tests discussed in this chapter to late-night worries about baby's first cold. Your pediatrician's office is an invaluable resource, especially if it offers an after-hours hotline allowing you to talk with a nurse about your concerns. In addition, the following sources can be handy to have on your bookshelf or bookmarked on your computer.

Books

Baby 411: Clear Answers and Smart Advice for Your Baby's First Year by Denise Fields and Ari Brown (2011)
Heading Home with Your Newborn: From Birth to Reality by Laura A. Jana and Jennifer Shu (2011)
Mama Doc Medicine by Wendy Sue Swanson (2014)

Websites

About Kids Health (website of The Hospital for Sick Children, Canada), http://www.aboutkidshealth.ca
After the Baby Arrives (from the CDC), http://www.CDC.gov/pregnancy/after.html
Evidence Based Birth (blog written by Rebecca Decker), http://evidencebasedbirth.com
HealthyChildren (from the AAP), www.healthychildren.org
Mayo Clinic (reliable website for general health questions), www.mayoclinic.org
Red Wine and Apple Sauce (blog written by Tara Haelle), http://www.redwineandapplesauce.com
Seattle Mama Doc (blog written by Wendy Sue Swanson), http://seattlemamadoc.seattlechildrens.org

rate and blood oxygen level. The test should be done at least 24 hours after birth and before discharge from the hospital.

The hepatitis B vaccine is also recommended for newborns within the first couple of days of birth. The rationale behind this vaccine is discussed in detail in appendix B.

The newborn screening tests and the hepatitis B vaccine detect or prevent rare, but potentially catastrophic, diseases or disorders. Their rarity can sometimes make parents think these procedures are unnecessary interventions, but understanding the science behind them reveals that without them, new babies would be at greater risk for serious outcomes. Weighing the risks and benefits, the brief discomfort or inconvenience of these procedures is easily outweighed by the peace of mind of knowing that your baby is protected.

4 ⁂ FOR ONCE, SIT BACK AND WATCH

How Newborns Explore, Communicate, and Connect

Baby Isla was born in a Toronto hospital in 2012. Like all human babies, the most neurologically immature of primates at birth, Isla was born with only 25% of her adult brain size.[1] She would depend on her parents to meet nearly all of her needs in the coming months. But beginning in the moments after her birth, she also demonstrated an extraordinary ability to explore her new world.

The doctor placed Isla on her mother's chest, and Isla looked up brightly at her face. Leah, her mom, recalled: "I kissed her and gave her the welcome-to-the-world speech I had prepared, and all the while she was making kissy lips at me. Then she started wriggling, found my nipple and started nursing. It was amazing. She knew just what to do."[2]

Over the course of about 20 minutes, baby Isla located her mom's breast, physically maneuvered to it, and latched on. I called this extraordinary, and it is, but it's also completely normal. If given the chance, many healthy newborns can find their way to their mothers' breasts in a similar way, a phenomenon called the "breast crawl."

A newborn baby goes through an incredible transition on his birth day. Within moments, he passes from a dark uterine home, the only one he's ever known, into a new world. His senses are bombarded with new sights, sounds, and smells. How does he navigate through this unfamiliar environment to his mother's breast? What makes him ready for this task?

This chapter was not part of my plan for this book. But as I read about the breast crawl and began to investigate these questions, I stumbled upon a rich body of scientific literature that seeks to understand how newborns sense the world. Most of our parenting questions, and much of this book, focus on what we should do to best care for our babies. This chapter is a bit different. It is about watching our babies, with wonder, and appreciating how they learn about the world—and us—from their first days of life.

THE BREAST CRAWL

The breast crawl has been carefully documented in research papers, including a 2011 study led by Swedish researcher Ann-Marie Widström.[3] The paper describes the behavior of 28 newborns placed on their mothers' chests, skin-to-skin, just after birth. In places, it reads like the narration of a wildlife documentary, recording the movements of the newborn human in his native habitat. "Gradually, the reflexes come to life," the story begins.

At birth, the babies in Widström's study cried for a couple of minutes as they were dried off and then placed on their mothers' chest. The mothers were encouraged to talk to their babies and to stroke them, but they were asked not to interfere with their movements. Instead, they watched as their brand new babies did the work. The babies soon grew quiet and began a characteristic sequence of behaviors:

> In the relaxation phase, the baby did not move any parts of the body, not even the mouth. Soon thereafter, the infant entered an awakening phase and started to make small thrusts with its head and small movements with its shoulders and arms. The awakening phase was followed by an active phase, when the infant showed more distinct activity. Most of the behaviors described . . . (e.g. looking at the breast, looking at the mother's face, rooting movements, hand-to-mouth activity, soliciting sounds) occurred during the active phase . . . During the crawling phase, the infant approached the areola [the darkly pigmented area surrounding the nipple], and during the familiarization phase, the baby became acquainted with the areola by licking and touching the

nipple before eventually entering the suckling phase, in which the baby started suckling the breast without help.[4]

Watch a baby working on the breast crawl, and you'll usually see a serene, focused face and deliberate movements. Kym, another mom who told me about her daughter's breast crawl, recalled: "There was no struggling or crying. I have pictures where the look on her face is just really intense and alert."[5] Indeed, among the careful research notations about babies attempting the breast crawl, there is no mention of babies getting fussy or frustrated in the process, although some of them do fall asleep on the job.

Widström and her colleagues believe that the breast crawl is a baby's chance to gradually, in his own time, become familiar with his environment and to naturally zero in on what is meant to be his source of nutrients for at least the next several months: mom's breasts. "The full-term healthy infant when skin-to-skin with its mother immediately after birth optimizes its ability to reach self-regulation within the first period of wakefulness when going through the inborn biological program to find the mother's breast," they write.[6]

But not all of the babies in Widström's breast crawl study began nursing on their own. All 28 babies actively looked for the breast, but only about two-thirds of them found it. And just over half began nursing on their own, on average an hour after birth. Infants that made more "soliciting sounds" (short, affirmative noises—not cries) and more hand movements from their mothers' breast to their own mouth were more likely to zero in on the areola and begin nursing.[7]

A lot of meaning and magic are attached to the breast crawl, and there are plenty of claims that it is a requisite part of the natural birth transition,[8] as if actual childbirth isn't enough of a rite of passage for a newborn baby. Although that makes a nice story, I couldn't find any evidence that babies who complete the breast crawl are better off than those cradled to their mom's chest and guided to the breast. No breast crawl study has followed babies for more than a day or so to see whether the successful crawlers end up as better feeders or more attached to their moms.

For Leah, watching baby Isla wiggle her way to the breast to begin her first feed gave her a huge sense of relief. She had been anxious about how breastfeeding would go, and seeing Isla take the lead in that first latch was reassuring. Of course, your baby might be different, and if you let your baby try the breast crawl, you shouldn't think of it as some kind of pass-or-fail test. Your baby might not find your breast. He might just rest on your chest. He might fumble around as if looking for something that is always out of reach. He might bob up and down on your nipple like a drunken sailor. He might very well need some help, and that's totally fine. He's a newborn baby after all, and you're his mother. You're in this together; you'll both play a part in getting to know each other in your own way.

All of that said, the fact that babies are born with a drive to search and find their source of nutrition within minutes of birth is impressive. A newborn baby can't stand up after a few minutes and wobble around to follow his mother like a newborn foal. He can't physically cling to his mother for protection, like the newborns of many other primate species.[9] And yet, he does have the ability to sleuth out a breast and to move toward it. What I like about the breast crawl is that it highlights how much newborn babies know at birth. And if we're comfortable pausing and letting them search for the breast, if we take this time to observe them, then we get to appreciate from the first day of life just how tuned into their world they are, through all their senses.

A NEWBORN'S SENSE OF TOUCH

Of all the senses, touch is the earliest to develop in fetal life. Beginning at around 8.5 weeks of gestation, touch receptors appear around the fetus's mouth, and then, between 10.5 and 12 weeks, around the genitals, palms of the hands, and soles of the feet. Finally, by 20 weeks, they spread throughout the surface of the skin. In newborns, a light touch in the palm of the hand activates more parts of the brain than a flash of light, the sound of a woman's voice, or a Chopin piano piece.[10] Touch is powerful.

Newborn babies exhibit a characteristic set of motor reflexes, involuntary movements that occur either spontaneously or in response to certain stimuli.[11] Many of these are dependent on touch receptors, and they're evaluated at birth to test for normal nerve and brain activity. For example, if you stroke a baby's cheek, he'll turn toward your touch. If you touch the roof of his mouth, he'll begin sucking. Both of these reflexes are essential to helping him find the nipple and begin feeding. Another reflex is grasping: stroke the palm of a baby's hand, and he will close his fingers around yours. The grasp reflex appears around 18 weeks of gestation, but the sucking reflex does

✲ NEWBORN REFLEXES

REFLEX NAME	*HOW TO ELICIT THE REFLEX*
Rooting reflex	Stroke the corner of the baby's mouth, and he will turn his head toward your touch, with mouth open.
Sucking reflex	Touch the roof of the baby's mouth, and he will begin sucking.
Moro reflex	Hold the baby facing up, supporting his head. Let the head drop slightly (1-2 cm) but suddenly. The baby will startle, arms and legs extending and fingers spreading, and then he'll pull his limbs back in. You may also observe this reflex (even in a sleeping baby) in response to a loud noise, sudden movement, or even the baby's own cry.
Grasping reflex (palmar and plantar)	Stroke the palm of the baby's hand, and he will close his fingers around yours (palmar grasp). Likewise, if you stroke the middle of the foot, from heel to toe, his toes will curl around your finger (plantar grasp).
Stepping reflex	Hold the baby upright with his feet touching a solid surface. He will move his feet in an alternating stepping motion.

Note: All of these reflexes indicate normal neurological development and are tested as part of the newborn exam.

not begin until week 32. This is why a very premature baby might be able to tightly grasp your finger but be unable to feed effectively on his own.[12]

The newborn reflexes may occur involuntarily, but they are not empty actions. For example, a newborn seems to be able to differentiate his own touch from that of another person, because the rooting reflex is much stronger when someone else touches a newborn's cheek than when the baby's hand brushes his own cheek.[13] He is also able to collect information about an object's shape and texture by exploring it with his hands and mouth. For example, if you place a small wooden cylinder in a baby's palm, he'll grasp it tightly, but if you present it to him repeatedly, he'll hold it for less and less time, presumably because he's losing interest in it. But if you then place in his hand a small wooden prism, with hard edges instead of rounded ones, he'll hold it for much longer, exploring the new shape with his sense of touch.[14] (This works both ways: whether you start with a cylinder or a prism, a baby will spend more time with the unfamiliar object.) This experiment has been repeated in premature babies as young as 28 weeks of gestation.[15] Even at this age, though underdeveloped in so many ways, babies are able to collect information about the world with their hands.

In another experiment, newborn babies were presented with two objects identical in size, shape, and texture. Both were small cylinders covered in the same soft rubber, but underneath, one was made of hard plastic and the other of squishy sponge. The cylinders were attached to pressure transducers that recorded the frequency and strength of the babies' "squeeze" on the objects. When these objects were placed, one at a time, in the babies' hands, they spent more time squeezing the hard object. But when the objects were gently placed in their mouth, they much preferred the squishy one, squeezing it more than the hard one.[16] Newborns' oral preference for soft things is probably helpful in getting started with feeding.

Babies are sensitive to touch, and as parents, we know this intuitively. Holding, stroking, patting—all of these gentle touches can communicate so much care to a baby. Touch can even help lessen a newborn's reaction to what is normally a painful procedure. In a randomized controlled trial, newborns held snugly skin-to-skin with their moms during a routine heel stick procedure (to collect blood for newborn screening) cried 82% less than those swaddled in a blanket in a bassinet.[17] Interestingly, this same study found that not all kinds of touch were soothing. The authors reported that when mothers stroked their newborn's head, the babies seemed irritated. As parents, we get lots of time to try different ways of soothing our babies, and we can learn to use the power of touch to comfort them as they adapt to the world.

WHAT DOES A NEWBORN HEAR?

A newborn baby knows his mother's voice. Even at 38 weeks, still in utero, a fetus can tell mom's voice from that of another woman. Play a recording of mom reciting a poem, and the baby's heart rate increases by five beats per minute. Play the sound of a strange woman reading the same poem, and his heart rate *decreases* by four beats per minute.[18] At birth, a baby prefers the sound of his native language to a foreign one,[19] as well as the sound of a familiar story—one that his mom recited many times during pregnancy—to a new one.[20]

In one experiment, newborn babies were given headphones and pacifiers that worked like little remote controls. If a baby sucked faster on the pacifier, it turned on a recording of his own mom reading Dr. Seuss's *And To Think That I Saw It on Mulberry Street*. Slower sucking turned on a recording of another woman reading the same story. In this study, 8 of 10 babies were able to figure out this system and preferentially use the pacifier to tune in to their own mother's voice.[21]

Newborns prefer familiar sounds, and these sounds may also help them to feel less stressed. In one study, 5-day-old babies were subjected to a heel stick. During the procedure, one group heard the sound of a maternal heartbeat, as it sounds in utero. Another heard the sound of a Japanese drum at the same volume and rhythm as the heartbeat, but with a different frequency of sound. The third group heard no sound. After the heel stick, all of the babies showed a rise in salivary cortisol, a marker of physiological stress, but the group that heard the maternal heartbeat had a much smaller rise in the stress hormone.[22] What is familiar is comforting. This isn't surprising, because we know it to be true for ourselves as well.

WHAT DOES A NEWBORN SEE?

Of all the senses, vision is the least developed at birth. Vision development requires maturation of different parts of the eye as well as integration with multiple parts of the brain, and this happens rapidly over the first year of life. But at birth, visual acuity, or clearness of vision, is not very good.[23] This is why newborns pay more attention to high-contrast, black-and-white images; they simply can't see more subtle gradations in brightness. Visual acuity is assessed in older children and adults with those familiar eye charts with letters of decreasing size. In infants, it is measured using cards with black lines of varying thickness, called the Teller Acuity Test.[24] A baby is shown a solid gray card with black-and-white stripes on either the left or the right side,

✲ NORMAL VISUAL DEVELOPMENT IN INFANTS

AGE	*CHARACTERISTICS OF VISUAL DEVELOPMENT*
Birth	Innate preference for human face
	Sensitive to bright lights; sees best in dim lighting
	Can resolve and pays attention to high contrast black-and-white patterns
	Will follow slow-moving, close objects or faces
	Able to see most colors
2 months	Holds eye contact and studies the features of your face
	Watches people from a distance
	Begins to study his own hands (called "hand regard"), marking early visual-motor integration
3 months	Shifts from auditory to visual dominance
	Seeks your attention by making eye contact and moving limbs
	Shows more interest in simple objects, though still prefers faces
4-6 months	Loves seeing the faces of other babies and own mirror image
	Shows recognition of familiar people on sight alone and smiles in response
	Sees an object of interest and will reach for and grasp it, then bring it closer for inspection
	Shows understanding that an object hidden from view still exists (object permanence)
6-12 months	Understands that a two-dimensional picture can represent an object
	Follows the direction of your gaze to share what you are seeing
	Reads the expression on your face to look for positive or negative responses to a situation
	Shows an understanding of connections between words and objects

Looks from an object to a parent to communicate wanting it

Shows a toy to a parent to share in the wonder of it

Source: P. Glass, "Development of the Visual System and Implications for Early Intervention," *Infants and Young Children* 15 (2002): 1-10.

and if he can distinguish the stripes, he'll preferentially gaze toward them. We know that at birth newborns can see high-contrast lines as thin as one-sixteenth of an inch, giving them a visual acuity of about 20/400 at birth. (For many children, their vision won't mature to 20/20 or better until 3 to 5 years of age.)[25] We also know that newborns can distinguish most colors at birth, although color vision improves as the baby develops.[26]

Toy companies have caught on to newborns' attention to high-contrast images and thus market lots of newborn-specific toys with bold black-and-white patterns, but there's no evidence that you need to provide this extra visual stimulation for your baby. Your home has plenty of natural visual stimulation: the stripes of window blinds, the rotation of a ceiling fan, and the edge of a doorway. And babies are even more interested in human faces, particularly when they come with interactions with caring adults. Your face will best hold your baby's attention if it's within 10 to 20 inches of his face, and dim lighting will encourage your baby to open his eyes and look around.[27]

In a classic 1975 study, infants just a few minutes old were shown several simple black-and-white images.[28] One had the features of a human face, two were scrambled versions (eyes at the bottom, mouth at the top, for example), and a blank version had just the outline of the face. The images were moved in an arc across the babies' line of sight. When they were shown the correct face image, the babies were more likely to turn their heads to follow it, compared with the scrambled or blank image. What is most extraordinary is that they preferred the image of a human face even though they hadn't seen a real one yet; all the people in the delivery room were decked out in facemasks, hair caps, and gowns. (It tells you something about standard practices in a 1970s delivery room that this experiment was conducted before the babies even met their mothers.) More recent studies have replicated the finding that newborns have an innate preference for the human face.[29]

Newborns also prefer a face that is socially engaged. If you show newborns two photos of the same woman, one with eyes open and one with eyes shut, they much prefer the one with eyes open.[30] They will also choose a face with a direct gaze, available for eye contact, over one with eyes averted to the side.[31] And newborns quickly learn to identify their own mother's face; within hours of birth, they recognize and prefer the face of mom to that of a stranger. However, recognition of mom's face is not just based on visual information; it also requires having heard her voice. In one experiment, mothers were asked not to talk to their newborns at all but to otherwise care for them as usual. (Can you imagine how unnatural this would feel?) Several hours after birth, these babies were shown photos of their mothers and a stranger, and they showed no preference for mom's face.[32]

These studies underscore the importance of eye contact and talking with our infants from the first day of their lives. And when you think about it, none of this is surprising. After all, it's how we humans best get to know one another: face-to-face, in conversation. Why should babies be any different? They're watching, listening, and making neural connections between the pattern of our faces and our familiar voices, known from the womb.

Because newborns pay attention to high-contrast images, there has also been some speculation that the natural darkening of the areola during pregnancy might help catch baby's attention and lead him to the nipple during the breast crawl.[33] In support of this idea, babies given a chance to do the breast crawl do spend a lot of time looking at the breast before they start moving toward it.[34] But nobody has tested whether the contrast of the areola affects the breast crawl, and while this idea sounds nice, it probably isn't true. That's what Dr. Penny Glass, a developmental psychologist at Children's National Medical Center, told me: "Given how relatively immature a newborn baby's vision actually is, particularly in relation to all the other sensory systems, I doubt that the contrast between the areola and the rest of a mom's breast is the primary stimulus that compels the baby to seek it."[35] Indeed, it is probably not sight (or sound) but a newborn's keen and discerning sense of smell that is the strongest sense guiding him along on the breast crawl.

WHAT DOES A NEWBORN BABY SMELL?

The importance of a newborn's sense of smell was demonstrated in an elegant study published in 1994.[36] Thirty moms and babies were included in the study. After each baby was born, one of the mother's breasts was washed and dried using an odorless soap. The baby was then placed on mom's chest, nose at midline and eyes at nipple level, and observed as he wiggled toward one or the other breast. All 30 of the babies found their chosen nipple on their own and made mouth contact with it, and just 5 of them required some help latching on. Which breast did the babies choose? Of the 30 babies, 22 chose the unwashed breast. There was a clear preference for the breast that carried mom's odors of her milk and body. The same test repeated at 3 to 4 days of age revealed the same preference, showing that the baby was still using mom's natural scent to find the breast.[37] However, the older babies had clearly honed their skills; it took them, on average, just 2 minutes of "crawling" to start nursing, whereas the newborns had needed an average of 50 minutes to begin to nurse.

Not surprisingly, newborn babies also like the smell of amniotic fluid, which they've been swimming in, tasting, and inhaling up until birth. If one of the mother's breasts is rubbed with amniotic fluid, most babies will crawl toward that breast rather than the untreated one.[38] In another study, the smell of amniotic fluid significantly decreased the amount of time newborns spent crying when separated from their moms.[39] By about the fifth day of life, babies begin to choose the natural-smelling breast over amniotic fluid.[40] At this age, a breastfed baby will also choose the smell of his own mom's breasts over a stranger's.[41] He's even attracted to the smell of his mother's armpits![42]

The odors and flavors in both amniotic fluid and breast milk are affected by the mother's diet. For example, when women eat garlic during pregnancy, their amniotic fluid (obtained by amniocentesis) is deemed garlicky by a panel of adult sniffers.[43] The same is true of breast milk from garlic-eating moms.[44] Babies probably shift their preference from amniotic fluid to breast milk a few days after birth because they're homing in on the specific flavors present in breast milk, which are associated with the satisfaction of a full belly.

You smell lovely to your newborn baby. This highly attuned sense of smell is surely part of what guides babies to their mother's breast just after birth and in the days that follow. Judging from how sensitive newborns are to smells, and how much they seek yours, your smell may help your baby feel comfortable and relaxed. To keep as many as possible of these olfactory cues intact, you might consider delaying bathing both yourself and your newborn for at least 24 hours after birth.

WHAT DOES A BABY LEARN IN THOSE FIRST MOMENTS WITH YOU?

At birth, your baby will turn toward your touch and the sound of your voice, will study the sight of the light and dark patterns of your face, and is drawn to the smells of your body. Biologically, he already knows that you are important. He wants to be close to you and to get to know you

⁎ BABY'S FIRST BATH CAN WAIT

Newborns are attracted to the scents of amniotic fluid and breast milk. Likewise, the sweet smell of her new baby triggers reward centers in mom's brain. After a long labor, you might really want a hot shower, and it may be hospital routine to bathe your baby right away. However, consider delaying both of your baths for at least the first day, so that you can get to know one another with all of these olfactory cues intact. It is enough to pat a baby dry and hold him skin-to-skin, covered in a warm blanket. Delaying the bath also allows the vernix, the cheesy white covering on a newborn's skin, to remain in place. Birth marks a rapid transition from the wet uterine environment to the relatively dry outside world, and vernix helps newborn skin stay hydrated as it adapts to the new climate. It also contains many proteins known to have antimicrobial and immune functions, so some researchers hypothesize that it might serve as a barrier to infection and/or encourage colonization of the skin by beneficial microbes.

J. A. Dyer, "Newborn Skin Care," *Seminars in Perinatology* 37 (2013): 3-7.

better. And when the two of you are close during this time, he is learning important things about you.

Part of what makes a newborn ready to learn is that he is very alert and aroused just after birth. He's left the nest of the warm, slow uterine world, where the placenta supplied constant nutrition. Now he needs to be able to breathe air and, with some help, find food. A healthy newborn shows that he is ready for this transition by looking sharp: having normal skin color, a strong heart rate, alert reflexes, active muscle tone, healthy breathing, and a vigorous cry. All of these aspects go into the Apgar score assigned to your newborn, first at one minute to indicate how well he tolerated the birth process and again at five minutes to see how he is adapting to the outside world.[45]

A newborn baby's state of alertness is probably related to the stress of vaginal birth, with all of those contractions and the tight passage through the cervix.[46] Beginning in the first stage of labor, a baby's stress hormones start to rise. By the time of birth, his blood concentration of norepinephrine is 20 times higher than prelabor levels. Norepinephrine is higher in babies that are born vaginally than those born by elective C-section, demonstrating that the physical trauma of the labor process—the squeeze and the brief periods of low oxygen during contractions—is probably important for initiating this normal stress response. By the time the baby opens his eyes to look around, hears your voice, takes a sniff of the world, and feels your skin, he's in a state of arousal. This is his moment. His norepinephrine level will stay high for the next couple of hours. And studies show that he is especially good at learning during this period, most likely because of his aroused state. After an hour or two, his stress hormone levels drop, and he'll usually enjoy a long, restorative sleep. During his alert period, your baby is actively collecting information about the world around him. One of the most interesting things that happen in the hour after birth goes back to the newborn's remarkable sense of smell.

A 2007 study investigated the olfactory learning of vaginally delivered babies by exposing them for 30 minutes to aroma oils in one of two flavors: cherry or pomegranate.[47] Half of the newborns were exposed to one of the smells during the first hour after birth. The rest were exposed to one of the smells 12 hours after birth. Then, a few days later, the babies were tested to see if they preferred one or the other scent. If, for example, cherry was what they smelled just after birth, which scent did they prefer several days later—cherry or pomegranate?

The aroma oils were dripped onto gauze pads that were then suspended on either side of each baby's head, and the babies were observed to see how much time they spent turned toward each pad. It turned out that how soon after birth the baby had experienced the fruity scent mattered a lot. The babies who got their sniff just after birth clearly preferred that same smell in the test several days later, as if it had made an impression on them. However, the babies that got their first whiff of the fruit scent 12 hours after birth showed no clear preference for one or the other smell in their test several days later. In another study, babies that had experienced spontaneous labor contractions before C-section delivery (versus a scheduled C-section) were better at the smell test and had higher norepinephrine levels.[48] The relationship between norepinephrine and olfactory learning in human babies is just a correlation, but studies in baby rats show that activation of norepinephrine receptors in the olfactory bulb of the brain is necessary for learning new smells.[49]

If the stress of birth primes babies to learn about their immediate surroundings, does a scheduled C-section mean your baby won't know you from the doctor? And what if, for whatever reason, you are unable to hold your baby in the first hour after birth? Before we jump to any judgment about these situations, let's remember that human babies have many ways of getting to know their parents. You are more than a 30-minute pomegranate exposure. Your baby will be savoring the smell of you—and your sight, sound, and touch—from your first day together and through the many, many days that follow.

GETTING TO KNOW YOU: IT GOES BOTH WAYS

We've marveled at how babies relate to their world and begin to learn about their important people, but what about us? We're using our senses to learn about our new babies as well. Several studies show that moms are really good, right off the bat, at identifying their new babies. For example, one study conducted a few days after birth found that mothers with their eyes and nose covered were able to pick out their own babies solely by stroking the back of their hands.[50] When the moms wore gloves, they were no longer able to identify their babies.

Most of us, before we become parents, think that all newborns look the same. But when we have our own babies, we see that this isn't the case at all. In another study, both moms and dads, after being with their new babies for at least 10 minutes, could almost always pick their own newborn's face from seven new-baby photos.[51] (To the dads reading this chapter, I apologize on behalf of the research community that you are otherwise so poorly represented in the studies on newborns. The dads really excelled in this last study. It was conducted in Israel in the 1980s, when fathers visited their new babies in the hospital only once a day, and they hadn't seen the baby for an average of 16 hours before taking this test. Fathers were just as good as the more involved mothers at identifying their babies.)

Sweet newborn baby smell draws us to our newborns as well. We don't know exactly why newborns smell so good. It could be amniotic fluid, the vernix coating their skin, or a combination of several chemical signals. Whatever it is, we like it. Researchers have used functional magnetic resonance imaging (fMRI) to measure brain activity in women while smelling the odor on a shirt worn by an unfamiliar 2-day-old infant. They found that the smell triggered activation of reward pathways mediated by dopamine in all of the women, but this activity was stronger among new moms in several areas of the brain.[52] Such signaling may help moms be more attuned to their own newborn's particular smell. In an older study, when moms were handed three onesies worn that day by newborns in the hospital nursery, 100% could pick out the shirt from their own baby.[53]

FROM BIOLOGY TO BOND

So far, the focus of this chapter has been biology: how our babies sense us and the world, how we sense them, and how these biological forces bring us together in the first days after birth. It is human attraction in its purest form. However, it's only the beginning. Building a real relationship with our babies is a much longer and more complex process, but observing and understanding your baby is the first step in building that relationship.

In the newborn stage, babies may seem to be simple eating, burping, crying, pooping aliens, but they are experiencing the world for the first time with intense sensitivity. That sensitivity allows them to learn at a rapid rate, but it can also be overwhelming, and newborns are easily overstimulated. Knowing this is a reminder that we don't need to add stimulation to a newborn's world. Your quiet, familiar voice is enough sound; your face, gazing at your baby, is enough to see; filtered light through a window is enough light and color; and resting against your chest is enough touch. And even this may be too much sometimes. Your baby might turn his face away from stimulation, and that's a sign that he needs a break from the sensory input. If you don't get that message, he'll probably use his most effective means of communication—crying. All of this is normal everyday stuff for a baby; what matters is how we respond.

You can use your knowledge of your baby's senses to sooth him and help him to regulate stress. For example, a randomized trial compared newborns' salivary cortisol as a measure of stress in response to two soothing methods. In one, the infants were given a gentle massage by a researcher trained in infant massage, but she didn't make eye contact with or talk to the baby. These babies had increased cortisol after the massage—they appeared to be *more* stressed. The second soothing method also included massage, but the researcher talked to the baby and interacted using eye contact throughout. She also adapted her touch to each infant's cues, moving on to massage a different part of the body if the baby seemed uncomfortable, for example. In these babies, salivary cortisol actually went down—they were *less* stressed.[54] Engaging your baby's senses in a gentle and responsive way can help him feel more comfortable in the world.

Observing your baby is key to understanding his needs and responding appropriately. Even everyday activities, like diaper changes, dressing, and bathing, can be stressful to a new baby, but you can mitigate this by showing sensitivity to your baby's signals.[55] One study, for example, scored maternal sensitivity and cooperation during a baby's bath. How well did mom observe and respond to the baby's signals? Did she follow the baby's pace and attention in the bath, or did she constantly begin new interactions and activities? The baby's salivary cortisol was measured after getting out of the bath as an indication of how stressful the transition was. All of the babies showed a rise in cortisol after the bath, but the ones whose moms were more sensitive and cooperative during bath time recovered to normal cortisol levels more quickly.[56]

This sensitivity to your baby—reading his cues and responding appropriately—takes some time and some trial and error to learn. And it depends a lot on the baby, too. Some babies are more reactive and need more help in regulating their physiological and emotional states than others. Regardless, sensitive observation and responsiveness can help you and your baby get through the day with less stress and more connection. It is also the foundation of attachment, your relationship with your baby that will build over the first year and beyond. A solid attachment relationship means that your baby knows you can be relied on when he needs you, and having this security, he is empowered to explore on his own as he grows, knowing that he can always come back to you. In modern parenting, we tend to talk a lot about specific practices that are supposed to build attachment, and these focus on physical closeness: breastfeeding, bed sharing, "baby wearing," and the like. There's actually little evidence that these parenting practices alone do much to build attachment,[57] but there is abundant evidence that parental sensitivity does.[58] You sooth your baby when he's upset; you feed him when he's hungry; and when he smiles at you, you smile back. This is how we listen to what babies are saying to us, in their own ways, and respond in the best ways we can to let them know that we care.

Babies are fascinating. They already know and can do so much when they come into our care, and they learn more every day from the world and from us. Just as much as it's a responsibility and a lot of work to care for them, it is a privilege to watch them grow. Sensitive observation, starting from the first day of life, rewards us with moments of marvel, a calmer baby, and a stronger relationship.

5 ⁂ MILK AND MOTHERHOOD

Breast Milk, Formula, and Feeding in the Real World

When Cee was handed to me just after birth, she came screaming and red-faced, with her eyes squinted shut. I said hello to her, and she stopped crying, opened her eyes wide, and gazed alertly into mine. And then, within a couple of minutes, she started moving her cheek against my breast, rooting for milk. I opened the hospital gown and held her clumsily, trying to remember the holds I'd practiced with a baby doll in my two hours of breastfeeding class a month before. A nurse confidently arranged a pillow under my arms and guided my hands in place. Cee did all the rest. She latched on and started nursing with the confidence of a pro. It was good that her instincts were so strong, because I'm not sure mine had kicked in yet.

I was determined to get everything right about motherhood, and feeding was no exception. I always planned to breastfeed, and between the two of us, Cee and I figured it out pretty quickly. After the first couple of weeks of nipple soreness and constant nursing, we settled into pleasant feeding routines. I loved this time with her, and it was empowering to know that my body could make this perfect food that could nourish her so completely. Breastfeeding was a big part of my identity as a new mother, and it was a source of pride. I relished the approval from my pediatrician, family, and friends, and I enjoyed the supportive glances from strangers. (I know many moms experience an overt *lack* of support when they breastfeed in public, so I consider myself lucky that I never did.) Because my experience was so positive, it was easy for me to be a little judgmental of women who didn't breastfeed, given the long list of benefits for both mother and baby.

Three years later, my brother and sister-in-law, Jordan and Cheryl Green, welcomed their own baby girl, Amy Bell. Cheryl planned to breastfeed and, like me, was surrounded by support, from Jordan, her grandmother, and her friends, among them lots of moms experienced with breastfeeding.[1] But beginning at the hospital, Cheryl's plans quickly unraveled. Amy Bell struggled to latch on correctly, and although she appeared to be feeding, her weight was dropping rapidly. Within her first couple of days of life, she lost 12% of her birth weight, and a lactation consultant urged Cheryl and Jordan to supplement with formula. For the next three weeks, Cheryl kept up a labor-intensive cycle of attempting to breastfeed, pumping, and supplementing with formula. Everyone—nurses, lactation consultants, and her friends—told her to keep trying, that it took time and practice, but still, Amy Bell didn't latch on, and very little milk came through the pump. Cheryl was scheduled to return to work at four weeks postpartum, and she didn't know how she would keep up these efforts on the job. Reluctantly, she and Jordan began exclusively feeding formula to Amy Bell.

Cheryl says she still feels a little guilty about not breastfeeding for longer, and she wonders if she missed out on a special bond with Amy Bell. But, she told me, it was also really helpful to be able to share feeding responsibilities with Jordan as they both learned the routines of new parenthood. For Jordan's part, he had been very attached to the idea of Cheryl breastfeeding their daughter. He grew up around breastfeeding, and he saw it as the normative and natural way for babies to be fed. But Jordan told me that he now appreciates that feeding, like all of parenting, is a "balance between ideals and practical realities." Thinking about Amy Bell, he said: "Now that I've watched her grow into an active, alert, engaged, and advanced baby, I feel confident that her needs are being met."[2]

Jordan is only bragging a little when he says that his daughter is advanced. Amy Bell is now 10 months old. It seems like she's hit nearly every milestone a little ahead of schedule, and she's never really been sick.[3] She and Cee are both beloved in our family, and nobody would ever think to wonder whether they'd been fed differently as babies.

Comparing my and Cheryl's breastfeeding stories, however, there is an impulse to call one a success and one a failure. That haunted me as I started working on this chapter. Cheryl's experience was riddled with challenges that I never had to face, and she tried harder than I ever had to. Her story of struggling to make enough milk is just as common as my happy story of breastfeeding for two years. And by most reasonable measures, Amy Bell and Cee are both big successes: They're happy, healthy, and well-nourished children, and both of our families have found our own ways of adjusting to new parenthood.

But for new mothers, it can be hard to find that perspective. Beginning in pregnancy (and often before), we all hear the same message: good mothers breastfeed—it's one of the most important gifts you can give your baby. This message translates into tremendous pressure to breastfeed, and we're quick to judge ourselves and each other if it doesn't work out. It is because of this pressure and judgment that how we feed our babies has become one of the battles in the "mommy wars." This is an unfortunate way to talk about feeding, one of the most important ways we care for our babies, whether by breast or by bottle.

Breastfeeding and its role in modern parenting is in part a story about science: how science has paved the way for good substitutes for breast milk while at the same time revealing the intricacies of breast milk, which no substitute is likely to replicate. But it's also about how science is translated to real life. How is it molded into public health messages intended to alter women's behavior? And what happens if breastfeeding, which should be the most natural way to feed babies, just doesn't work?

A SHORT HISTORY OF THE SCIENCE OF INFANT FEEDING

The ability to make milk to feed our young is what makes us mammals, and as humans, we evolved to produce a milk uniquely suited to meeting the nutritional and immunological needs of human babies. Breastfeeding is the biological norm, and it is how the majority of young infants have been fed throughout most of the history of our species.

There have always been substitutes for breastfeeding, though, and following their history is a fascinating way to follow the science of milk. For a long time, there was no science to guide infant feeding strategies; mothers and other caregivers just pieced together what they could. If a mother didn't make enough milk, had to work away from home, or died in childbirth, or if a baby had an oral handicap that impeded nursing, then other options were needed. Sometimes this meant another lactating woman, maybe a family member or friend, would help nurse the baby, and sometimes a wet nurse was hired expressly for this purpose. Records of wet nurses go back at least as far as the third or fourth century BC.[4]

But if human milk wasn't available, substitutes were used. Since wet nurses were being paid to feed another woman's baby, sometimes their own babies would be denied enough milk from their moms and would need these substitutes.[5] Almost as soon as cows and other dairy animals were domesticated, their milk was used for infants, sometimes placing babies directly on the teat to nurse.[6] Infant feeding vessels have been found in children's graves throughout the Roman Empire, dating back to 4000 BC.[7] By the 1400s, soon after the invention of the printing press, printed books offered advice and recipes for homemade supplements called pap or panada. These usually contained a cooked combination of several ingredients, including cow's or goat's milk, bread crumbs, flour, meat broth, honey, egg, and sometimes even wine or beer.[8] These concoctions could be used as the primary food for a baby or as a supplement to breast milk. Cross-cultural historical records indicate that two-thirds of preindustrialized societies introduced some solid foods to babies before 6 months of age, sometimes as early as a few weeks of life.[9]

Throughout most of history, it was probably self-evident that substitutes were inferior to breast milk and often resulted in illness. Ironically, this situation became especially dire in the eighteenth and nineteenth centuries, when it was a common belief that boiling cow's milk made it less nutritious. Raw milk was usually swimming in bacteria by the time it traveled, unrefrigerated, from farm to baby. During this time, babies fed breast milk substitutes suffered and died disproportionately from diarrhea, particularly during the summer months. In the late 1800s, nearly all bottle-fed infants in New York City orphanages died.[10]

Enter science. In the late 1800s, Louis Pasteur's work showed that bacteria caused disease and that they could be killed with pasteurization. Water chlorination and modern sewage systems meant clean water for feeding and for cleaning bottles and nipples. By the early 1900s, the availability of kitchen iceboxes and canned evaporated milk meant that relatively safe formulas could be made at home.[11]

The study of nutrition was also exploding. By the late 1800s, scientists understood that not all milks are alike. Cow's milk has more protein and less sugar than human milk, so scientists and pediatricians began recommending recipes meant to be a closer match. A common recipe that could be made at home called for one 13-ounce can of evaporated cow's milk, 19 ounces of water, and 1 ounce of Karo corn syrup. Scurvy and rickets were common problems, but by the 1920s, supplementation with fruit or vegetable juice and cod liver oil decreased the incidence of these vitamin deficiency diseases.[12]

As science revealed more and more about nutrition, the recommended formula recipes grew more complex. Food companies stepped in to offer commercial products, relieving hospitals, institutions, and moms of having to make their own and creating a huge, profitable market. By the 1950s, commercial formulas had gained popularity and began to replace homemade recipes.[13] These products were, for the most part, nutritionally adequate, clean, and consistent. For the first time in human history, babies could be exclusively fed a breast milk substitute without a noticeable risk to their health. Most parents and pediatricians assumed that formula was just as good as, if not better (being more "scientific") than, breast milk. Mothers increasingly turned to doctors for advice, and doctors recommended that breastfeeding moms feed their infants on a schedule, typically every four hours. If that didn't seem to satisfy the baby, then supplementation with formula was needed.[14]

Other societal changes made formula feeding the preferred choice for modern women. By the mid-1900s, most women were giving birth in hospitals, where they were separated from their babies soon after birth and allowed only brief, scheduled visits for feeding, making it difficult to establish breastfeeding.[15] But women were also looking to break free of their duties as full-time mother and housewife. Particularly during World War II, formula allowed women to fill important jobs in the workforce, and after the war, they didn't want to give up their careers.[16] Breastfeeding went from necessary to optional to out of style. By 1970, it had reached an all-time low: only one in four infants were breastfed past one week of age.[17]

But around the same time, women began fighting for more freedom from medical authority in childbirth and parenting, and a renewed appreciation for breastfeeding was part of this movement.[18] Scientists, meanwhile, were beginning to take a closer look at breast milk and were finding that it was much more than just a collection of nutrients. While formulas based on cow's milk or soy can be made to contain a similar amount of protein, fat, and carbohydrate, these nutrients are of better quality and more easily digested in breast milk than in formulas.[19] Breast milk also provides a dynamic suite of immunological proteins, growth factors, stem cells, digestive enzymes, hormones, and prebiotics.[20] We can now appreciate that breast milk probably evolved to include many of these components because they're good for babies, and investigating health outcomes in breastfed and formula-fed babies has been a very active area of research for the past several decades.

The history of breast milk substitutes is a reminder that they've always been needed, but only in very recent human history has science allowed for a safe alternative. That there is even a debate over breast versus bottle is made possible by science. It's also fueled by the science examining potential benefits of breastfeeding. This science, however, is difficult to do and even harder to interpret in a meaningful way.

HOW DO WE STUDY BREASTFEEDING?

By the 1980s, there was increasing research interest in comparing health outcomes in breastfed and formula-fed babies, but this area of science has been challenging from the start. If we wanted to test the hypothesis that breastfed babies are smarter, for example, the best way to do this would be with a randomized controlled trial. We would randomly assign one group of brand-new moms to breastfeed and one group to formula-feed, and then we'd test cognitive development later in childhood. But this is a problematic study design for obvious reasons. Breastfeeding is not just a medical intervention. It's a behavior that requires coordinated effort between mother and baby, and it is helped or hindered by a range of biological, cultural, and economic realities. Women's choices about feeding their babies are complex, and any researcher hoping to randomly dictate how a baby should be fed is bound to be disappointed. It would also be unethical, because the research we have to date indicates a range of health benefits from breastfeeding. Conducting a true randomized trial of breastfeeding is pretty much impossible.

Instead, most breastfeeding research is based on observational data. To investigate the question of cognitive development, for example, researchers can compare test scores in children who were breastfed and those who were formula-fed. But this kind of information, on its own, really isn't useful, because breastfed and formula-fed babies may be different in lots of other ways besides the type of milk they drank.

This is one of the major challenges to breastfeeding research. For example, a 2006 study of childhood intelligence and breastfeeding found that mothers who breastfed had higher IQ and more education and provided a more stimulating and supportive home environment than moms who formula-fed.[21] Breastfeeding moms were also more likely to be white and older and less likely to be poor or to smoke. These disparities demonstrate that, in a society with such a huge gap between rich and poor, breastfeeding is often a privilege. Children who are breastfed are more likely to be born into families in which moms can afford to stay at home or work in the types of jobs that give them paid maternity leave, on-site child care, and the flexibility to pump breast milk when needed during the workday. These disparities can confound our interpretation of the data, so we call them confounding factors. The breastfed children in the study just described did score 4% to 5% higher on the intelligence tests, but was that because of the breast milk or because they were born into so many other advantages?

Observational studies deal with confounding factors by including them in their statistical analysis, in effect trying to mathematically level the playing field for babies. The study just described did an excellent job of this, and when the researchers accounted for all of those confounding factors, it turned out that the breastfed kids scored just about 0.5% higher than the formula-fed kids, but this difference was no longer significant. (That is, the effect was so small and variable that, statistically, it wasn't different from zero.)[22] This is just an example; other studies of cognitive development and breastfeeding do find a more convincing benefit, which I discuss later.

Even the best observational studies can fall short, since breastfeeding is so strongly stratified by social class. We can never really know whether we've accounted for every confounding factor. And yet, the vast majority of what we think we know about the benefits of breastfeeding is based on observational data. Many of these studies, particularly the older ones, do a poor job of controlling for confounding factors, so we always have to interpret their findings with care and a healthy dose of skepticism.[23] In my review of the literature for this chapter, I've paid attention only to studies that made a solid attempt at adjusting for relevant confounding factors.

BENEFITS OF BREASTFEEDING: THINGS WE FEEL PRETTY SURE ABOUT

There is good evidence that breastfeeding protects babies from gastrointestinal, ear, and respiratory tract infections—important benefits because these are the most common causes of illness in infancy.[24] This protection is especially important in developing countries, where lack of clean water and access to medical care means more babies get sick and fewer get medical attention, but it is relevant in developed countries as well. The degree of protection provided by breastfeeding in the developed world varies among studies, but breastfed infants seem to have a 25% to 75% reduction in infections,[25] with better protection associated with exclusive breastfeeding.[26] This effect isn't found in all studies, and confounding factors can still be an issue. For example, many studies don't consider whether or not babies are in day care or have siblings, factors that can affect exposure to pathogens and can be related to feeding choices. However, studies that do a solid job of adjusting for confounders consistently find that breastfed infants are less likely to come down with these infections. This effect seems to be limited to the period of breastfeeding, however, with protection quickly waning after weaning from breast milk.[27]

One reason that we feel pretty certain about breastfeeding's role in reducing infections is that there's a clear biological mechanism for it, given that breast milk has a range of immunological properties. Breast milk contains immunoglobulins that allow mom to passively pass on her immunity (gained from infection or vaccination) to her baby. Oligosaccharides bind up pathogenic bacteria in the intestine, preventing infection, while also promoting the growth of beneficial bacteria. Various other components nonspecifically attack bacterial and viral pathogens, inhibit bacterial growth, modulate inflammation, and stimulate the infant's developing immune system.[28]

The *act* of breastfeeding might also protect babies from ear infections. Breastfeeding requires the baby to create strong pressure in a rhythmic suck, swallow, breathe pattern, and this is thought to keep the Eustachian tube in the inner ear aerated. Bottle feeding (whether breast milk or formula) creates less pressure, which can result in the pooling of milk in the Eustachian tubes and middle ear.[29] Bottle feeding in a semi-upright position can help prevent milk pooling and might reduce the risk of ear infections in bottle-fed babies.[30]

We are also fairly certain that breastfeeding protects babies from SIDS. A meta-analysis of 18 studies found that any amount of breastfeeding reduces the risk of SIDS by about half, and at least two months of exclusive breastfeeding further reduces risk.[31] This may be explained by breastfeeding's protection against infections, since babies that die of SIDS often have minor bacterial or viral infections before death.[32] Experiments have also shown that formula-fed babies are more difficult to arouse from sleep, which might increase their susceptibility to SIDS.[33]

A few other benefits of breastfeeding are also worth mentioning. There is strong evidence that feeding premature babies with human milk rather than formula is associated with a significant decrease in the incidence of necrotizing enterocolitis (NEC), a life-threatening condition in which the intestinal tissue dies.[34] There's also some evidence that breastfeeding slightly reduces the rates of childhood leukemia, although a mechanism for this is uncertain.[35]

WHAT WE'RE NOT SO SURE ABOUT: LONG-TERM BENEFITS OF BREASTFEEDING

In 2014, Ohio State University sociologists Cynthia Colen and David Ramey published a study with the provocative title "Is Breast Truly Best?"[36] It was a well-designed study including more than 8,000 children and looking at long-term outcomes, including body mass index (BMI), obesity, asthma, parental attachment, hyperactivity, and a range of cognitive tests. When Colen and Ramey analyzed their data across families, incorporating statistical adjustment for confounding factors—including socioeconomic status, maternal education, prenatal care, and preterm birth, among others—they found that breastfeeding appeared to be beneficial for all of the outcomes tested except asthma (for which breastfeeding seemed to increase risk). However, they then limited their analysis to a group of nearly 1,800 siblings who were fed differently as babies, some breastfed and some formula-fed, and everything changed about their results: breastfeeding seemed to have no impact on any of the measured outcomes. In other words, when they looked only at kids who were raised in the same home, by the same mother, whether they were breastfed or formula-fed mattered little to how they turned out later in childhood. This sibling cohort study design is powerful because it nearly eliminates the problem of confounding factors.

The release of Colen and Ramey's study caused a big stir. Media outlets, always happy to fan the flames of the mommy wars, crowed: "Breastfeeding benefits have been drastically overstated" and "Breast milk is no better for a baby than bottled milk."[37] Formula-feeding moms celebrated that their kids were not doomed to obesity and general subpar performance in life. Breastfeeding advocates picked apart the methods of the study with a rigor rarely applied to studies finding benefits of breastfeeding and worried that these new findings might derail breastfeeding promotion efforts.[38] The frenzy over this one study shows just how personal science can feel when it relates to our own parenting practices or to a cause in which we are invested.

Colen and Ramey's study was a big deal because it contradicted public health messages that have told us, over and over, that breastfed children have a life-long advantage over formula-fed children. The AAP's policy statement on breastfeeding, written by the organization's Section on Breastfeeding, lists improved cognitive development and prevention of obesity and asthma as long-term benefits of breastfeeding.[39] Michelle Obama's obesity prevention initiative, Let's Move!, includes breastfeeding as one of the first steps you can take to prevent your child from becoming obese.[40]

But in fact, for those who had followed previous research on the long-term effects of breastfeeding, Colen and Ramey's findings weren't at all surprising. Outcomes like obesity, intelligence, and asthma are measured years after a baby ingested any breast milk, and a lot of other factors affect them in the meantime. Previous studies have found conflicting results for all of these outcomes. Almost everyone seemed to miss the point that this was just one study among many to evaluate the long-term effects of breastfeeding, and it didn't detract from the well-established short-term benefits that we've already discussed.

One of the best studies to look at long-term outcomes from infant feeding is the Promotion of Breastfeeding Intervention Trial (PROBIT), led by Michael Kramer of McGill University.[41] In this study, 31 maternity hospitals and linked pediatric clinics in the Republic of Belarus were randomly assigned either to begin promoting and supporting breastfeeding using the WHO/UNICEF Baby-Friendly Hospital model (the intervention group) or to continue with their usual practices, which weren't particularly helpful for breastfeeding (the control group). This made PROBIT a randomized controlled trial without directly telling women how to feed their babies. PROBIT included more than 17,000 babies born in 1996-97, and follow-up studies have so far tracked them through 11 years of age. Although all women in the PROBIT study began breastfeeding after the birth of their babies, the intervention led to big increases in breastfeeding duration and exclusivity. For example, 43% of the intervention moms were exclusively breastfeeding at three months compared with only 6% of the control moms. Because the intervention was randomized, the two populations of families were otherwise the same, allowing PROBIT to look at infant feeding outcomes on a large, public health scale.

The PROBIT study found that babies in the intervention group, with more breastfeeding, had fewer bouts of diarrhea and lower rates of eczema in the first year of life (although there was no difference in rates of ear or respiratory infection).[42] However, when the study looked at the children at 6.5 years of age, they found no difference in BMI, obesity, blood pressure, asthma, behavioral difficulties, or rates of dental cavities.[43] And again, at 11 years of age, there was no difference in BMI, obesity, or risk factors for heart or metabolic disease.[44] However, at age 6.5, the study found a small increase in verbal IQ and in teachers' ratings of children's reading and writing abilities in the intervention group.[45]

As the closest thing we have to a randomized study of breastfeeding, the PROBIT study made huge contributions to our understanding of possible long-term outcomes of infant feeding. In short, it shows, just as Colen and Ramey did, that the benefits of breastfeeding are mostly limited to infancy. The one notable exception is for cognitive development, and a few other high-quality studies support this idea. A Harvard study that did a good job of controlling for maternal IQ and the quality of the home environment found that longer duration of breastfeeding was associated with higher language scores at age 3 years and greater IQ at age 7, adding up to about a 4-point (out of 100) increase with 12 months of breastfeeding.[46] Another study found that breastfed babies in both Britain and Brazil had a small but significant increase in IQ. Looking at both of these countries helped to reduce the effects of confounding factors, because breastfeeding is associated with higher education and socioeconomic status in Britain but is equally common across all social classes in Brazil.[47] Although Colen and Ramey, as well as another sibling cohort study, found no effect of breastfeeding on IQ, a third sibling study did.[48]

Taken together, this evidence suggests that breastfeeding might cause a small improvement in infants' cognitive development. This could be due to long-chain polyunsaturated fatty acids, especially docosahexaenoic acid (DHA), which are necessary for brain and vision development and are present in breast milk but were only recently added to infant formulas.[49] It could be that hormonal differences in breastfeeding moms, such as increased oxytocin and prolactin, might help them be more relaxed and focused on their babies, and the greater amount of contact between mom and baby during breastfeeding might lead to more social stimulation that could enhance cognitive development.[50]

What about obesity? Are formula-fed kids fatter? There is very little evidence that this is the case. The PROBIT study found no effect of the breastfeeding intervention on later obesity or risk factors for heart disease.[51] In addition to Colen and Ramey's study, another sibling cohort study found no difference in obesity later in childhood.[52] Observational studies that adjust for potentially confounding factors find little to no effect.[53] Studies that compare outcomes across cultures with different social patterns of breastfeeding find no consistent effect of breastfeeding on obesity or risk of type 2 diabetes.[54] And really, all of this isn't surprising. So many factors influence obesity and later metabolic health. In the United States, where social advantages greatly increase a baby's chances of being breastfed, breastfed babies are also probably more likely to have access to healthy food and safe places to play later in childhood. Obesity is a disease of inequality, and it seems more likely that breastfeeding is a marker of that inequality rather than a cause of it.

The data on breastfeeding and asthma are interesting. Many studies have found that breastfed babies are less likely to have wheezing or asthma in the first couple of years of life.[55] However, in studies that look at later childhood, this effect disappears and sometimes even reverses. The PROBIT study found no difference in rates of asthma or allergies between the breastfeeding intervention and control groups when children were 6.5 years old.[56] A large, prospective study in Tucson, Arizona, found that infants who were exclusively breastfed had lower rates of wheezing in the first three years of life. However, by age 13, children that had asthmatic mothers and were exclusively breastfed for at least four months had a much higher incidence of asthma—nearly ninefold higher than children who weren't breastfed.[57] (For mothers without asthma, breastfeeding had no effect on the child's risk of developing asthma.) Studies in New Zealand and Australia have found similar results: breastfeeding may decrease the risk of asthma early in life, but later in childhood and adulthood, the risk is increased.[58] Researchers think that the early protection against wheezing may be related to the reduction in respiratory illnesses, which might cause wheezing and other asthma-like symptoms in babies and young children (but this isn't true asthma).[59] The later increased risk could be related to components in the milk of asthmatic mothers, the use of asthma medication, or changes in infant microbial and pathogen exposure, or it could be related to study bias and confounding.[60] We don't know. Regardless, breastfeeding doesn't seem to confer a long-term benefit in reducing the incidence of asthma.

I want to briefly mention maternal benefits to breastfeeding. Breastfeeding is associated with a lower risk of breast and ovarian cancer, maybe because it lowers a woman's lifetime exposure to estrogen. Some, but not all, studies have found that breastfeeding is associated with a lower risk of developing diabetes and heart disease.[61] However, all of the data on long-term health outcomes are observational, so they are limited by the same confounding factors as long-term infant outcomes.

I know I just bombarded you with a lot of science. Sorry about that, but I think it is important to be clear about what the science does and doesn't tell us, because this can have an impact on our feeding choices and how we feel about them. After several decades of research on breastfeeding outcomes, here, in summary, is what we know. During infancy, breastfeeding reduces a baby's risk of gastrointestinal, ear, and respiratory infections, as well as SIDS. Later in childhood, kids that were breastfed may have a small increase in IQ. There are lots of other proposed benefits, but they're most likely related to confounding factors, because the most rigorous studies don't actually support them. I will also add that there's a ton we don't know about breast milk. It's still a relatively mysterious biological substance, and lots of cool studies on this are emerging every year. New research may reveal more benefits of breastfeeding, and it may also allow some improvements to infant formulas.

SCIENCE, TRUTH, AND GOOD INTENTIONS GONE WRONG

After sorting through all of this science on breastfeeding benefits, I was struck by how messy and full of conflicting results it is. The scientists writing these reports are usually honest about the limitations of their work and careful in its interpretation, but somewhere between these studies and their translation into mass media, public policy, and public health campaigns, this honesty and nuance seems to get lost.

We can see this in the 2012 report "Breastfeeding and the Use of Human Milk," written by the AAP's Section on Breastfeeding, which lists a multitude of benefits of breastfeeding with little acknowledgment of the conflicting science behind them.[62] For example, it sounds quite certain that breastfeeding lowers a child's risk of asthma: "There is a protective effect of exclusive breastfeeding for 3 to 4 months in reducing the incidence of clinical asthma, atopic dermatitis, and eczema by 27% in a low-risk population and up to 42% in infants with positive family history." Contrast this to the 2008 report on infant nutrition and atopic disease, written by the AAP's Committee on Nutrition and the Section on Allergy and Immunology, which gives a much more measured and accurate review of the research.[63] On asthma, it says: "In summary, at the present time, it is not possible to conclude that exclusive breastfeeding protects young children who are at risk of atopic disease from developing asthma in the long term (>6 years), and it may even have a detrimental effect." Why are these two wings of the AAP telling such different versions of the science?

This gets to the crux of the problem with breastfeeding science and policy. The science is hard to do well and hard to interpret. Breastfeeding advocates, who clearly make up the AAP's Section on Breastfeeding, have made it their life's work to encourage more women to breastfeed, and they must feel it is warranted to smooth over the wrinkles in the data for the good of that goal. These are our nation's breastfeeding experts, and we'd like to think that they're our best source of information, but it turns out that their bias appears to prevent them from giving us the straight story on the science.

I'm not the only one who is disappointed in the obvious bias in breastfeeding information. In a paper entitled "The Problem with Breastfeeding Discourse," sociologist Dr. Stephanie Knaak wrote: "Mothers empower health professionals by seeking out their knowledge and expertise. Health professionals, in turn, empower mothers by respecting their decision-making autonomy. The lynchpin of this relationship is the communication of scientifically-sound, impartial information. When the information becomes biased, the discourse takes on a manipulative character, threatening the foundation of trust so central to this relationship."[64]

We do science because we want to understand the truth. It is a constant, ongoing pursuit to understand our own biology so that we can make smart choices at an individual and policy level. If the science of breastfeeding is used first and foremost as a tool for breastfeeding promotion, we compromise public trust in science. Biased information about breastfeeding also sets up infant feeding as a debate, which sometimes escalates to mommy war status, and it doesn't need to be either of these. If moms are told that breastfeeding will solve the woes of a chronically ill society (along with the message that every mom can breastfeed), then of course moms who breastfeed can become judgmental of those who don't, and of course formula-feeding moms can end up feeling ashamed and defensive. The funny thing is that if we want to encourage more women to breastfeed, we have plenty of solid data to back its benefits. If we focus instead on the real benefits of breastfeeding while also acknowledging the real barriers and difficulties that women face, then there is no debate. For many families, breastfeeding is a good choice. For others, it isn't, and we can move on to talk about how to support all new moms and their babies. But honesty is essential in how we talk about the science of and the barriers to breastfeeding; otherwise we unnecessarily increase the heavy burden of meeting the ideals of motherhood.

IF BREAST IS BEST, WHY DOESN'T EVERY MOM DO IT?

Breastfeeding is good for babies, and science backs substantial benefits. Why, then, do so few women meet the AAP's recommendation to breastfeed exclusively for six months, continuing, with the addition of solid foods, through at least the baby's first birthday?[65] In 2010, although 77% of U.S. moms initiated breastfeeding at birth, only 16% exclusively breastfed to 6 months, and just 27% were still breastfeeding to some extent at 12 months.[66] These rates are a vast improvement over those of just a generation or two ago, but they are a far cry from the goals that the AAP set for us.[67]

In the United States, part of the problem is that we aren't the most family-friendly country in the world. We are one of only four countries (of 173) to have no national policy requiring paid maternity leave.[68] It wasn't until 2010 that, under the Affordable Care Act, employers were required to provide breaks and a clean and private space (i.e., not a bathroom) for breastfeeding employees to pump.[69] However, this law applies only to employers with 50 or more employees, and pumping breaks are unpaid, so they extend a mother's workday and time away from her children. Lack of economic and societal support for breastfeeding in the United States makes it awfully hard to breastfeed exclusively for six months, as much as many moms might love to do so.

But looking around the world, we can see that this isn't the whole story. Take Norway, for example. Norway doesn't allow formula advertising or freebies. Nationally, every family has up to 46 weeks of fully paid parental leave or 56 weeks with 80% pay, and each parent is also allowed to take up to another year of unpaid leave. Although 99% of Norwegian moms initiate breastfeeding at birth, less than 50% exclusively breastfeed to four months and less than 10% make it to six months.[70] So, even with model societal support for breastfeeding, these women aren't meeting the standards set by major health organizations.

If it were just about the health benefits of breast milk, I suspect that most mothers would breastfeed and breastfeed for longer durations. But clearly, it isn't that simple. Some moms, like Cheryl, just don't make enough milk. In the United States, research indicates that 10% to 20% of women struggle to make enough milk in the first few weeks postpartum. In a Colorado study of first-time moms who were motivated to breastfeed and were provided with good lactation support, 15% were unable to make enough milk to support their infants' growth with exclusive breastfeeding after three weeks.[71] A California study found that 19% of exclusively breastfed infants lost more than 10% of their body weight in the first three days of life, and for 42% of mothers, their milk hadn't "come in" by this time.[72] This amount of weight loss is associated with severe jaundice and dehydration, and nearly everyone agrees that it is an indication that the baby needs supplementation and the mom and baby need help with breastfeeding.[73]

Breastfeeding difficulties often persist beyond the immediate postpartum period, despite women's efforts to establish a full milk supply. A study led by Dr. Alison Stuebe of the University of North Carolina found that one in eight women (13%) had what the researchers called "disrupted lactation," meaning that they reported at least two of the following reasons for weaning: pain, low milk supply, and difficulty with infant latch. In this group of women, two out of three asked for help with breastfeeding from a health professional, but only one-quarter of them said that the help was actually helpful. Those with disrupted lactation breastfed for an average of five weeks, compared with seven months for the rest of the women in this study.[74]

There's some evidence that the magnitude of this problem may be unique to the United States, or at least to Western developed countries. Maybe it is because we're all relearning to breastfeed after a few generations of not doing so, and we're learning it in an often disjointed medical system, in which obstetricians focus on the mom, pediatricians on the baby, and lactation consultants on the breasts. We also know that delivering by cesarean, being a first-time mom, and maternal diabetes, obesity, stress, and older age are all associated with delays in milk coming in, and these factors have increased in the United States over the past generation or two.[75] Studies in Peru and Ghana have found that in these settings, unlike the United States, for the vast majority of women, their milk comes in within three days of birth, probably making for an easier start at breastfeeding.[76]

Some of the women who end up having the hardest time with breastfeeding are among those who are the most vulnerable during the postpartum period: those who have symptoms of postpartum depression. Another study by Alison Stuebe showed that women who are depressed or anxious during pregnancy and in the weeks following birth have lower oxytocin levels and release less oxytocin during breastfeeding, which could affect the ability to establish and enjoy breastfeeding.[77] While many women report that breastfeeding makes them feel more relaxed and less stressed,[78] new moms struggling with depression report feeling *more* overwhelmed, stressed, and depressed during a feed.[79] Stuebe believes that the neuroendocrine mechanisms underlying depression, on top of all the hormonal shifts that occur around childbirth, predispose women at risk for depression to also be at risk for breastfeeding problems.[80] Indeed, many studies find that feelings of depression and anxiety in the postpartum period are associated with breastfeeding problems and early weaning.[81] For a new mom who plans to breastfeed and wants desperately to do so, feeling like she's failing at what is supposed to be the natural way to feed her baby can be devastating, particularly against the backdrop of postpartum depression. Stuebe urges health care providers to be aware of this connection and to look beyond breastfeeding as the end goal: "If, for this mother, and this baby, extracting milk and delivering it to her infant have overshadowed all other aspects of their relationship, it may be that exclusive breastfeeding is not best for them—in fact, it may not even be good for them."[82]

Survivors of childhood sexual abuse, thought to number 20% of women in North America,[83] are also at risk for breastfeeding difficulties. Some find that breastfeeding is a way to reclaim their bodies and identities as women, but others find breastfeeding disturbing and may experience panic attacks and flashbacks of abuse.[84] Women with a history of eating disorders, approximately 13% of women in the United States, may also struggle with breastfeeding, grappling with issues of body image, loss of control, and anxiety about attention on their bodies.[85] This shouldn't discourage these women, or those with a history of depression, from breastfeeding. Breastfeeding can still work beautifully.[86] But these same women should also know that it might not go well, that this too is normal, and that a happy and healthy mom is much more valuable to her baby than any amount of breast milk.

While medical studies have focused on why we should breastfeed and how to get more women to do it, sociologists have been interested in hearing women's stories and learning how they feel about their infant feeding experiences. Their most compelling finding is that women's experiences are highly variable and rarely what the women expected. For some, like me, learning to breastfeed is a relatively straightforward process, and feeding is a pleasant experience of connection with the baby.[87] However, negative experiences are more common; in one study, two out of three women interviewed described breastfeeding as distorting, disrupting, painful, difficult, or disconnecting.[88] Most women, regardless of their breastfeeding experience, agreed that it was harder than they thought it would be.[89] Because they weren't prepared for this, those who faced serious difficulties establishing breastfeeding tended to blame themselves and reported feelings of failure and helplessness.[90]

These are stories that aren't being told in prenatal breastfeeding classes or promotion materials. Instead, the picture we are given is of a natural, mutually beneficial process, one in which mother and baby exist in harmony, each feeding an opportunity for an infusion of both health and love.

"Breastfeeding is the most natural, healthy, and loving gift a mother can give to her baby," reads one state health department's web page.[91] Another calls breastfeeding "the gift of feeling safe and loved."[92] When they mention challenges, they offer quick reassurance that they can be solved with enough persistence and support.

It seems to me that there's a missed opportunity to educate women about breastfeeding challenges in prenatal classes and other promotion

✲ RECOMMENDED INFANT FEEDING RESOURCES

Books

Bottled Up: How the Way We Feed Babies Has Come to Define Motherhood, and Why It Shouldn't by Suzanne Barston (2012)

Child of Mine: Feeding with Love and Good Sense by Ellyn Satter (2000). (This book covers feeding from infancy through preschool, including both breastfeeding and formula feeding.)

Great Expectations: The Essential Guide to Breastfeeding by Marian Neifert (2009)

The Nursing Mother's Companion by Kathleen Huggins (2010)

Websites

Bottle Babies (bottle feeding support website), www.bottlebabies.org

Fearless Formula Feeder (blog written by Suzanne Barston), www.fearlessformulafeeder.com

Getting Started with Breastfeeding (website from Stanford School of Medicine), http://newborns.stanford.edu/Breastfeeding

HealthyChildren (website for parents from the AAP), www.healthychildren.org

LactMed (a database of drugs and chemicals with information about safety for breastfeeding; also offers a smartphone app), http://toxnet.nlm.nih.gov/newtoxnet/lactmed.htm

The Leaky Boob (breastfeeding support website and blog), http://theleakyboob.com

activities. The focus is on convincing us to breastfeed and on boosting our confidence, and perhaps advocates fear that more realistic information might be demoralizing. But new mothers are also highly motivated to provide the best for their babies, and it would be better to prepare them for these challenges before they become emotional, exhausted, vulnerable new moms, rather than blind-siding them with these difficulties later. Research supports this more honest approach: women who have realistic expectations about breastfeeding are more likely to persevere through difficulties than women who are taken off-guard by them.[93] At the very least, this type of breastfeeding education would help struggling moms to feel less alone and would foster more empathy and a more caring dialogue between women regardless of their ultimate feeding experiences.

FEEDING IS MORE THAN JUST FOOD

Feeding is love. It is one of the primary jobs of parenthood. In the first few weeks and months of a new baby's life, we feed her every few hours, day and night. It is one of the first, and constantly repeated, ways in which we communicate to our babies that we are here to meet their needs. By responding to our babies' cues, we show them that we're listening to what they have to say and that we can be counted upon to care for them with love and respect. This is the foundation of attachment, whether your baby is breastfed or fed expressed milk or formula from a bottle.[94]

My formula-fed niece, Amy Bell, knows this is true. Jordan, my brother, told me about a time a few months ago when he and his daughter attended a meet-and-greet event for local political candidates at a neighborhood park.[95] Jordan fed Amy Bell a bottle, and after she'd finished her milk, she started to drift off to sleep with the contentment of a full belly and in the safety of her dad's arms. An older woman, probably about the age of Amy Bell's grandmothers, told Jordan that it almost looked like they could be breastfeeding, the way Amy Bell was snuggled so closely to his chest. He took that as a compliment. Was that bottle of formula the same as a breastfeed? Of course not. But to Amy Bell, it conveyed the most important messages of love, trust, comfort, and connection.

This is ultimately what I most want to tell new moms about infant feeding. Breastfeeding is important because it improves infants' health, but for an individual mom and baby, this is just one factor that might influence their ultimate experience and choice. I cherish the memories of breastfeeding my daughter, and I want every mother to have that same opportunity. I believe breastfeeding is a reproductive right, and we should be supported in every way possible to establish breastfeeding and to feed our babies wherever and whenever we want to. I also believe, and science supports, that women face a diverse array of very real barriers to breastfeeding. We need more research and better support for these women, and we also need to support mothers and babies when formula feeding is required or chosen. We need to appreciate each feeding—by breast or bottle—as a chance to build attachment and connection with our babies and to know that this is the greatest gift we can give them. The early days of parenthood are hard, and feeding is important, no matter what kind of milk the baby is drinking. We all deserve to be cheered on for the work we do.

6 ✻ WHERE SHOULD YOUR BABY SLEEP?

Sleep Safety and the Bed-Sharing Debate

Cee was 3 weeks old, and I felt like I was being swallowed into a fog of cumulative sleep deprivation. My mother had returned to her own life several thousand miles away, and my husband was flying around the country for job interviews. I was facing newborn night duty on my own, and between diaper changes and breastfeeding and soothing to sleep, I was often piecing together my nights in 30-minute increments. One night, after I'd nursed Cee in bed in the wee hours, we fell asleep together. I woke up to find sunlight streaming across the bed and my baby still sleeping sweetly next to me, her eyelids newborn translucent.

"Crap!" I thought. "We're cosleeping. I don't want to cosleep!"

But it happened again the next night and the night after that. When my husband returned home, he was relegated to sleeping on the couch, while Cee and I shared our double bed.

With Cee in my bed, I was feeling more rested, and I no longer resented her frequent night awakenings. Sleeping next to her made it easier to breastfeed during the night, and she often went back to sleep without a fuss when she could feel my body close to hers.

Still, I was uneasy about sleeping with Cee. I worried that I might roll onto her or pull a blanket over her head during the night. With this at the back of my mind, I slept fitfully next to her, fretting about the position of her small body and mine. I would often wake to feel my heart pounding like a jackhammer, and the only thing that would calm it was seeing the gentle rise and fall of Cee's chest as she breathed peacefully in her sleep.

I called my mom to get her advice. Besides being my mother and a wise woman in general, my mom worked for a social service agency supporting mothers and babies, so she talked about babies' sleep as part of her job.

"Alice, the safest place for Cee to sleep is in her own bed."

My mother's voice was gentle and empathetic, but it was firm. She was right, I thought. I put Cee back in her bassinet next to my bed after our nighttime feedings. I was more comfortable with this arrangement, and it turned out that Cee adjusted easily to sleeping on her own. Our flirtation with cosleeping was over.

Cosleeping wasn't a good fit for me, but many parents feel differently. A few months after Cee was born, my friend Esmee gave birth to a little boy named Miller. Miller slept in the crook of Esmee's arm on the night of his birth. Home from the hospital, he joined his parents in their bed, and two years later, they still can't imagine it any other way. Esmee believes that the safest place for her baby to sleep is right next to her, and she doesn't care what her mother or anyone else has to say about the matter.[1]

Even now, as 3-year-old Cee sleeps in her own big girl's bed, I am still fascinated with the debate around where babies should sleep. Peek into bedrooms around the world, and you'll find babies spending the night snuggled against their mamas, often breastfeeding on demand through the night. You'll also find babies sleeping in cribs in their own rooms, many going for long stretches without parental contact. Can we say which one is better? I dove into the research to find out. This chapter focuses on sleep safety and breastfeeding. I discuss how bed sharing affects sleep patterns in chapter 7.

Before I go on, let me clarify the vocabulary around this topic. The term *cosleeping* means different things to different people. To avoid confusion, I will use the term *bed sharing* to specifically mean that the baby and at least one parent sleep in the same bed and *room sharing* to mean that they sleep in the same room.

INFANT SLEEP IN CULTURAL CONTEXT

We can't talk about infant sleep practices without first putting them into cultural context. After all, cribs are a relatively recent invention in the history of human parenting. Our evolutionary foremothers probably slept with their babies out of necessity, to keep them warm and safe. In modern times, most moms and babies around the world still sleep in close proximity—in the same bed, or at least the same room.

For example, among Mayan families in rural Guatemala, it is normal for a mother to share her bed with her youngest child, breastfeeding as needed through the night.[2] After two or three years, a new baby is born, and the toddler moves into a bed with dad or siblings. When researchers told these Mayan families that babies in the United States often sleep alone, they were horrified, and, the researchers wrote, "Most of the families regarded their sleeping arrangements as the only reasonable way for a baby and parents to sleep." In these families, nobody slept alone. Even a widowed grandmother or single auntie is likely to have a child in her bed.

On the other side of the world, in a more industrialized and technologically advanced society, most Japanese babies sleep with their mothers as well. Fathers may work long hours and sleep in a separate room away from the rest of the family. The sleep habits of Japanese families reflect their priorities for relationships. Historically, Japanese culture has valued the mother-child bond, as opposed to the husband-wife bond, as the "most intimate family relationship."[3]

In the United States and other Western countries, the cultural norm over the past several generations has been for babies to sleep in cribs. Mainstream parenting books and magazines usually advise that babies sleep in a bassinet in their parents' bedroom for the first few months but then move to their own bedroom. Western cultures value a baby's ability to fall asleep independently and to sleep through the night on his own. In many families, if bed sharing occurs, it is not because parents want to sleep with their children but because they can't afford a crib or have exhausted other options for coping with sleep problems.[4]

But today, more and more Western parents are choosing to bed-share.[5] In some families, it's cultural; bed sharing is more common in African American, Hispanic, and Asian families in the United States.[6] An increasing number of parents, like Esmee and her husband, are choosing to bed-share as part of their parenting philosophy, believing it will build a closer bond with their baby. Others do it, not out of any kind of philosophical conviction, but just because they find that it works to get their family more sleep, and few things matter more during the early days and weeks of parenting.

When did parents start putting babies to sleep in cribs? There isn't a clear answer to this question, but scholars have identified several possible reasons for this shift.[7] For example, there are horrifying accounts, going back 500 years, of poor mothers in urban Europe confessing to priests that they intentionally smothered their babies in bed because they simply couldn't afford to feed all of their children. Lacking reliable birth control methods, this was the only way they could control their family size. The Church responded by campaigning against bed sharing, and some countries banned it. In the following centuries, there was an increased emphasis on the concept of "romantic love" and a greater desire for parental privacy. And in the early twentieth century, Freud was sounding the alarm that exposing infants to sex might cause lasting psychological harm. Having multiple bedrooms in a family home was also seen as an indicator of status and wealth. Finally, there was the belief that infants that slept alone would develop greater independence and self-reliance, attributes highly valued in industrialized Western nations.

While these factors may have contributed to the Western cultural shift in infant sleep practices, they seem outdated to today's parents. There is no evidence, for example, to support the idea that babies that sleep alone become more independent and self-reliant adult members of society. To my mind, the one serious objection to bed sharing that remains is the concern that it isn't safe.

CONFLICTING SAFETY ADVICE

When my mom told me that Cee wasn't safe sleeping with me, she was only repeating the stance of the American Academy of Pediatrics,[8] as well as other public safety organizations.[9] This recommendation isn't an arbitrary judgment on what should go on in our bedrooms at night; it is based on many studies showing that bed sharing is associated with an increased risk of infant deaths from SIDS and accidents such as suffocation.

Some researchers and parenting advice authors disagree with this recommendation. They say that bed sharing is the natural way for mothers and babies to sleep, that it facilitates bonding and breastfeeding, and that it can be done safely. For example, the website AskDrSears, written by Dr. Bill Sears and family, states that bed sharing is safer than putting your baby to sleep in a crib.[10] Notre Dame anthropologist James McKenna is a staunch advocate of bed sharing, and he's openly critical of those who discourage it. "The public wars being led by governmental agencies and by medical groups against bed sharing are nothing less than disrespectful and vitriolic of parents who choose to bed share safely," he wrote.[11]

Faced with this contradictory advice, you and I are left to make our own decisions on the long and lonely nights of caring for a baby. We all want to keep our babies safe, but we might also find the idea of sleeping next to them appealing and even necessary. We can't help but wonder, how good exactly are the data behind the recommendations against bed sharing?

Before I tackle this question, I want to warn you that this is not an easy topic to discuss. Whatever the reasons and risk factors for sudden infant deaths, each is a tragedy to the families that survive them. It is my intent to understand these deaths so that we can do our best to prevent them, not to assign blame or to scare you into one or the other choice.

SUDDEN INFANT DEATHS AND THE SLEEP ENVIRONMENT

Sudden infant death syndrome (SIDS) and accidental suffocation, asphyxia, and entrapment are the most common causes of death between 1 week and 1 year of age in developed countries.[12] SIDS and accidental deaths can be difficult to tell apart even after an autopsy and scene investigation, and many of their risk factors are similar.[13] According to CDC criteria, a diagnosis of suffocation or asphyxia requires evidence that obstruction of the baby's breathing occurred.[14] An infant death is diagnosed as SIDS when the cause of death cannot be determined, even after an autopsy and scene investigation. But to make things more confusing, a case can also be labeled as "undetermined." For example, a medical examiner might call a death undetermined if the scene investigation suggests suffocation but the autopsy does not confirm it. Under this classification system, about 30% of sudden unexpected infant deaths in the United States are classified as SIDS, 30% as suffocation, and 40% as undetermined.[15]

Studies show that both SIDS and accidental deaths such as asphyxia occur more frequently when babies bed-share, although they certainly happen in cribs as well. I'll discuss SIDS in detail later in the chapter, but let's briefly consider accidental deaths. Major causes of accidental infant deaths during sleep include an adult rolling onto the baby, soft bedding blocking the baby's airway, and the baby becoming entrapped, such as between a mattress and a headboard or the slats of a crib. In a U.S. study published in 2012, only 13% of infants that died in their sleep of suffocation were sleeping in a crib or bassinet.[16] Fifty-two percent were in an adult bed, 19% were on a couch or chair, and most were sleeping with an adult at the time of death. Those numbers make solo sleep in a crib look relatively safe compared with sharing sleep on a surface not designed for babies.

Of course, there are dangerous cribs in the world—cribs that are old, broken, or poorly designed, or those containing loose bedding, pillows, or bumper pads. However, in a crib that meets modern safety codes and contains just a tight-fitting sheet, there just aren't that many possible hazards to the baby. The crib is a controlled, almost sterile environment. An adult bed has many more factors in play: bedding, pillows, headboard, not to mention a parent or two and maybe another child.

With bed sharing, reducing suffocation hazards means addressing both the bed itself and the sleeping partner. The safest bed-sharing environment is probably a firm mattress on the floor, with no bed frame, and little or nothing in the way of blankets and pillows that might cover a baby's head. One of the most dangerous places for a parent and infant to sleep together is on a couch or chair. Any sleep partners should be unimpaired by drugs or alcohol and should not be particularly heavy sleepers. Nonparent adults, children, and pets are less likely to be aware of the baby during the night and should not bed-share.[17] We can't guarantee that sleeping with a much bigger person doesn't pose some risk to the baby, even in ideal circumstances, but taking these steps can go a long way toward lowering the risk of suffocation and asphyxia. SIDS is scarier, though, because it doesn't have an obvious cause, making prevention a daunting task.

UNDERSTANDING SIDS

We don't know exactly what happens when a baby dies of SIDS, but for some reason, these babies stop breathing in their sleep. Most babies will wake up if they feel they are struggling to breathe, but babies that die of SIDS don't.[18] Studies have found that many SIDS victims had abnormalities in the brainstem area associated with autonomic control of breathing, sleep, and arousal.[19] We can hope that science may some day give us the tools to identify these most vulnerable babies early on so that parents can do their best to protect them.

In the meantime, the working model explaining SIDS is the Triple Risk Model,[20] which states that for SIDS to occur, three types of risk usually come together:

1. *Critical period of development:* By definition, SIDS occurs in babies less than 12 months of age, but most deaths occur before 6 months of age, a period of rapid maturation of cardiorespiratory control and sleep-wake cycles.[21]
2. *Intrinsic risks:* Some babies are born more vulnerable to SIDS. Intrinsic risks include being premature, small for gestational age, or exposed to cigarette smoking during pregnancy, or having genetic defects such as the brainstem abnormalities mentioned above.[22]
3. *Extrinsic stressors:* These include prone sleeping (on tummy), overbundling, face covering, soft or excessive bedding, and mild upper respiratory infection. Bed sharing is thought to be one of these extrinsic stressors.

The Triple Risk Model shows that, in most cases, SIDS victims have multiple risk factors. A recent study of SIDS deaths in San Diego, California, between 1991 and 2008 found that 75% of SIDS victims had at least one intrinsic and one extrinsic risk factor.[23] This means that a baby could have an underlying brainstem abnormality, but if he always sleeps in a safe sleep environment, he will probably remain safe from SIDS. As parents, understanding these risk factors and minimizing the number of factors present for our babies can go a long way toward reducing the risk of SIDS.

Case-control studies are used to identify risk factors for SIDS. In a case-control study, researchers monitor SIDS deaths as they occur. When an infant dies of SIDS, several infants that were born around the same time in the same population are recruited as controls. Parents and caregivers of both the case infants (those who died of SIDS) and the control infants are interviewed. Were the babies exposed to smoking during pregnancy or after birth? Were they sleeping on their backs or tummies? Where were they sleeping when they died? Were they sleeping alone or with another person? Did their parents drink or use drugs? All of these factors are tallied up for the case and control infants, and then statistical analyses determine differences between the two groups. The results are presented as odds ratios, which give the odds of SIDS occurring in babies with the factor of interest. The higher the odds ratio, the greater the risk associated with that particular factor.

Case-control studies have some serious limitations, though. Most importantly, they can show which factors are associated with SIDS deaths but can't establish causation. In addition, they rely on parents' memory and honesty for information. For example, in their interviews with researchers, parents might decide to leave out the fact that they had a couple of drinks or more before bed. Or they just might not remember what exactly happened on the night in question. Parents of the babies who died of SIDS may be more likely to remember details, because the horror of finding your baby dead might intensify memories around the event. (This difference in remembering is called recall bias.) In other words, case-control data are only as accurate as parents are honest and accurate.

A randomized controlled trial would give us a more accurate picture of the role of a factor like bed sharing in SIDS, but it would be unethical and nearly impossible to do. We can't randomize babies by sleep habits. That would mean instructing some parents to bed-share and some to put their babies in cribs, and as you can imagine, it is unlikely that parents would follow those instructions, given differences in parenting styles and infant temperament. Another barrier to this kind of study is that it would have to be huge. Since only about 1 in 2,000 infants die of SIDS, we'd need to study hundreds of thousands of families to detect a difference in SIDS rates between bed-sharing and crib-sleeping infants. This leaves us with case-control studies. Imperfect as they are, they are the best tool we have for studying population-wide risk factors associated with SIDS.

A BRIEF HISTORY OF SIDS RESEARCH AND ADVICE

In the 1970s and 1980s, SIDS rates were at an all-time high in Western countries, and researchers turned to case-control studies to try to figure out why. One extrinsic risk factor that popped up over and over was the prone sleep position (sleeping stomach down). Babies that died of SIDS were much more likely to have been placed to sleep on their tummies or to have rolled there from the side position than the surviving controls. In the early 1990s, Back-to-Sleep campaigns urged parents to put their babies down for sleep on their backs, and within a couple of years, SIDS rates had plummeted.[24]

This bit of SIDS history illustrates "modern" parenting advice gone wrong. In 1955, Dr. Benjamin Spock recommended in his book *Baby and Child Care* that babies be placed to sleep on their backs. This was customary in most of the world at the time and probably throughout human history, because parents feared their babies would suffocate on their tummies. However, in his 1956 edition, Spock changed his tune and told parents to put their babies to sleep on their tummies, citing concerns that a baby sleeping on his back might choke on his own vomit.[25]

Tragically, nothing about Spock's advice was evidence-based. There weren't any good studies to support his ideas. But prone sleep seemed to work great, since babies sleep longer and deeper on their tummies,[26] and Spock's advice was quickly adopted by parents. Unfortunately, we now know that prone sleep decreases that critical ability to arouse from sleep, putting babies at greater risk for SIDS.[27]

Dr. Spock couldn't have known how harmful his advice would be. He wasn't alone in recommending prone sleep, but he had such a huge audience that he does bear some responsibility for this mistake. His book was printed in 42 languages and sold 50 million copies.[28] By 1970, there were two case-control studies pointing to prone sleep as a risk factor for SIDS, but Spock continued to recommend tummy sleeping until 1978. A 2005 study estimated that at least 60,000 infant deaths could have been prevented had the Back-to-Sleep campaign started in 1974.[29] Of course, the campaign didn't begin until the early 1990s, so Spock wasn't the only one to be a bit slow in educating parents about the importance of sleep position.

I include the story of Dr. Spock and prone sleep for two reasons: to illustrate the utility of case-control studies and as a cautionary tale. The case-control studies—with all their flaws—are what alerted us to the dangers of prone sleep. And it's a cautionary story in that Spock gave the advice that seemed logical to him, with tragic results. To me, this is a reminder that we should not ignore human traditions in parenting—including supine infant sleep and sleeping close to our babies. We should also take care to be sure that our recommendations are backed by solid science.

BED SHARING AND SIDS

The success of Back-to-Sleep encouraged researchers to look for more factors that might be putting babies at risk for SIDS. Since the 1980s, epidemiologists in Europe, New Zealand, and the United States have investigated thousands of SIDS deaths in case-control studies, trying to understand why some babies die of SIDS while others do not.

Overwhelmingly, these studies find that babies that die of SIDS are more likely to have been bed sharing than the surviving (control) babies. But they also show that infant sleep environments are complex, and we can't look at any one factor in isolation. In other words, maybe it isn't bed sharing per se that is dangerous, but rather the circumstances in which we bed-share—factors that we could change to minimize the risk of SIDS.

For example, smoking (by the expectant mother during pregnancy or anyone in the household after birth) clearly puts babies at risk for SIDS, and bed sharing further increases that risk. In a large study led by U.K. epidemiologist Bob Carpenter and conducted across 20 regions of Europe, smoking without bed sharing roughly doubled the odds of SIDS. When babies were exposed to both smoking and bed sharing, they had a 17-fold increase in risk of dying of SIDS compared with a baby with neither factor.[30] Suffice it to say that if your baby has been or is exposed to cigarette smoke, you should not bed-share. Period. Everyone agrees on this.

In the same European study, when the researchers looked just at the nonsmoking families, they found no increased risk of SIDS associated with bed sharing.[31] Many other studies have come to similar conclusions.[32] Other factors that increase the risk of bed sharing include parents' use of alcohol and drugs, sleeping on a couch, parental fatigue, loose bedding and pillows, and sleeping with premature or low birth weight babies. When parents avoid sleeping with their babies in these conditions, the risk of SIDS doesn't seem to be increased by bed sharing.[33]

But there's an important—and controversial—caveat to this conclusion. Several studies have looked separately at young babies, during the peak period for SIDS. They find that bed sharing does increase the risk of SIDS in babies younger than 3 or 4 months, even in nonsmoking families.[34] Babies that die of SIDS while bed sharing also tend to be younger, averaging 7 to 9 weeks of age, than those who die in cribs, at an average age of 15 to 16 weeks.[35] This potential risk to young infants is problematic because many parents begin bed sharing during the newborn period, whether intentionally or because it is the only way they can get any sleep.

As I was working on this chapter, a new study investigating this question was published, again led by Bob Carpenter.[36] The lead researchers of five major case-control studies from Europe and New Zealand pooled their data, giving a total of 1,472 SIDS babies and 4,679 controls. With this large number of cases, they specifically wanted to find out whether bed sharing increased the SIDS risk if babies were breastfed, parents didn't smoke, and the mother didn't use alcohol or drugs. The answer was yes, but again, only in young infants. In infants less than 15 weeks old, bed sharing in the absence of these other risk factors increased the risk of dying of SIDS fivefold. In infants older than 15 weeks, bed sharing did not increase the SIDS risk. Carpenter and his colleagues concluded: "Our findings suggest that professionals and the literature should take a more definite stand against bed sharing, especially for babies under three months."

Not surprisingly, the media blasted this news of the dangers of bed sharing around the world. And not surprisingly, breastfeeding and bed sharing advocates blasted back with critiques of the study.[37] For example, the alcohol and drug use data weren't complete or accurate, the study didn't consider the effects of heavy bedding, and it didn't differentiate between partial and exclusive breastfeeding. In other words, Carpenter and his colleagues didn't do a great job of quantifying some important confounding factors, and those who value bed sharing were quick to point this out and to reassure parents that careful bed sharing with young infants is just fine.

As with any paper published in a reputable journal, Carpenter's new study was peer-reviewed by other experts in the field before publication. And, lucky for us, it was published in *BMJ Open*, which makes peer reviews available to the public—a level of transparency that is rare in scientific publishing. Carpenter's paper had four peer reviews, all by leading SIDS researchers.[38] Three of them were very positive. The fourth was extremely critical. Dr. Peter Blair, another U.K. researcher who has published several case-control studies of SIDS, wrote this last review.

In Blair's review and Carpenter's subsequent response, there is a noticeable tension. Carpenter (like most SIDS researchers) thinks bed sharing is a serious risk to young infants, and Blair doesn't buy it. Both researchers have dug in their heels, and reading their words, you get the feeling that they may never agree. And they're both expert statisticians, so their arguments are written in statistical mumbo-jumbo that most of us can't follow. These two experts can look at the same sets of numbers and come up with vastly different conclusions.

A year later, just as my book was going to press (I added this paragraph at the last minute!), Peter Blair and his colleagues published their own paper.[39] Like Carpenter's paper, this one combined data from historical studies, but it did a better job of controlling for alcohol use. Blair and his coauthors found that bed sharing did not increase the risk of SIDS, even in infants younger than 3 months, when they excluded SIDS cases that involved parents' excessive alcohol use (more than two drinks), smoking, or sleeping with the infant on a sofa or chair (all conditions that are quite dangerous). This study was well-designed and well-controlled, and the data were really reassuring for parents. But still, it is just one study, and we can't ignore the previous studies with different findings.

All of this is frustrating for parents. The science is confusing and imperfect. It feels like scientists in this area have gone beyond their role as scientists and become advocates, and so we have to worry a bit about whether they are able to look at this question without bias. That they disagree is good for science because it leads to more rigorous studies. However, we can expect this issue to continue to be controversial, with more conflicting studies being released in the coming years. In the meantime, while researchers argue back and forth, parents are left wondering whom to believe and, ultimately, might distrust one side or the other or, worse, distrust science in general.

The truth is that the science is always evolving, and we may never know for sure whether bed sharing with young babies poses an increased risk of SIDS. Case-control studies will always have limitations, and if you go looking for holes in them, you can find them. Most SIDS experts and the AAP choose to err on the side of caution and recommend avoiding bed sharing for at least the first three to four months of life. Still, for whatever reason, parents will choose to bed-share. Maybe it is an integral part of their culture or their parenting philosophy, or maybe their baby simply won't sleep alone. In that last case, the risk of being up with a sleepless baby night after night and suffering the effects of chronic, severe sleep deprivation (think postpartum depression, car accidents, falling asleep in exhaustion with the baby on a couch, etc.) might easily outweigh the relatively small risk of bed sharing in an otherwise ideal situation. I think it is really important to talk about the safety concerns but also to recognize that "banning" bed sharing comes across as being out of touch with the reality of parenting and could also get in the way of open conversations about how to make the bed sharing environment safer.

Toward that end, it might be helpful to consider some possible explanations for the risks of bed sharing with young babies. A few studies have looked closely, using infrared cameras and physiological monitors, at the behavior of sober, breastfeeding moms sleeping with their babies in their own homes. They have found that it is common for bed-sharing babies to end up with their faces covered by blankets during the night and, as a consequence, to have brief periods of low oxygen and high carbon dioxide. For example, among 40 bed-sharing infants observed for two consecutive nights, 22 of them had a total of 102 incidents of head covering by blankets, most of the incidents caused by a parent's movement, and some babies slept like this for hours. In contrast, there was only one incident of head covering in a matched group of 40 babies who slept in cribs.[40] Another study found that of 20 infants observed for one night of bed sharing, 7 of them had a parent's limb resting on them for part of the night.[41] All of these incidents occurred on nights when parents knew they were being videotaped, so they may have been more careful than usual.

Clearly, bed sharing introduces more variability into the environment and increases the chances that a baby's breathing might be challenged by a change in oxygen availability during the night. In these sleep studies, all the babies were able to respond to small respiratory challenges by increasing their breathing and heart rates or by moving or crying to wake a parent. It is probably the intrinsically vulnerable infants—like those exposed to cigarette smoke, born prematurely, or having an unrecognized brainstem abnormality (unrepresented in these small studies)—who might fail to arouse and correct the problem.

Since we can't usually know in advance which babies are vulnerable due to something like a genetic defect, it seems prudent to arrange a baby's sleep environment to prevent head covering as much as possible. Putting your baby to sleep in a crib without blankets is one way to do that. Another option is to use a sidecar cosleeper (such as the one made by Arm's Reach) that attaches to the side of your bed and provides your baby with a separate sleep surface, away from your blankets and limbs. If your baby bed-shares with you, eliminate or at least minimize blankets and arrange them in such a way that they are never close to him. The baby can wear a sleep sack to keep him warm without the risk of covering his face. Whatever your interpretation of the science and wherever your baby ultimately sleeps, I think everyone agrees that making the sleep environment as safe as possible should be a priority.

THE CASE FOR KEEPING YOUR BABY CLOSE

In the 1980s, British pediatrician D. P. Davies was working in Hong Kong, where he was surprised to find that SIDS was virtually absent from childhood mortality statistics. Davies surveyed forensic pathologists and pediatricians at the major hospitals in Hong Kong; all agreed that SIDS was very uncommon.[42] They closely tracked babies that died of SIDS during 12 months in 1986-87, finding a total of 21 deaths, approximately one-fifth the rate in the United States at that time.[43] What were these Hong Kong families doing right, or more to the point, what were American families doing wrong?

Life wasn't all rosy for a baby in Hong Kong in the 1980s. Breastfeeding was rare, living conditions were extremely crowded, and rates of respiratory infection were high. However, smoking and prone sleep were uncommon, and most babies were born into supportive extended families. Davies speculated: "I wonder whether there might be some benefit to such high-density living. Babies are left alone much less. Sleep patterns might be different, effecting subtle modulations to physiological responses concerned with ventilatory control. The question 'When can I put baby into his own room?' is virtually never raised. Might closer overall contact with the sleeping baby somehow lessen the risks of sudden death?"[44]

Similar observations have been made in other cultures. A survey of nearly 5,000 families in 17 countries found that bed sharing and room sharing were common in many, but not all, countries with low SIDS rates.[45] Close sleeping arrangements are also common in Asian immigrant communities in the United Kingdom and among Pacific Islanders in New Zealand. These groups have low rates of SIDS, particularly compared with their neighbors of European descent.[46]

From these cross-cultural data, we can only speculate as to the role of close sleeping in SIDS rates. It may be other things about the way babies are raised, like prone sleep or lack of smoking, that could explain these

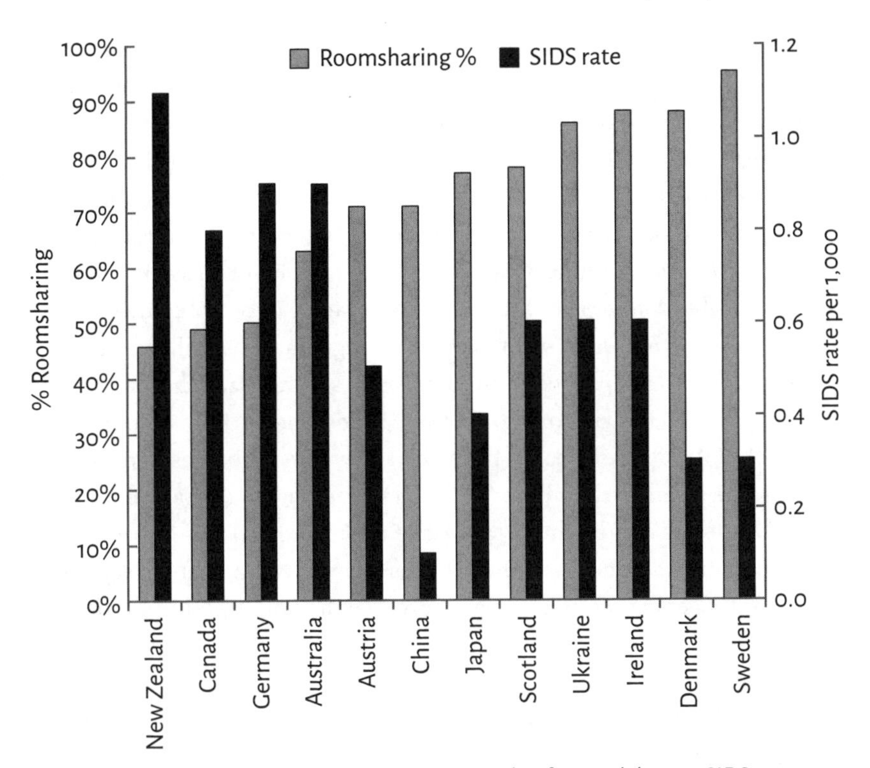

International survey of room sharing rates at 3 months of age and the 1995 SIDS rate (per 1,000 infants). Some of the lowest SIDS rates are observed in countries where room sharing is common, although many other variables also contribute to the SIDS rate. *Source:* E. A. Nelson et al., "International Child Care Practices Study: Infant Sleeping Environment," *Early Human Development* 62 (2001): 43-55.

observations. On the other hand, if bed sharing were truly a major independent risk factor for SIDS, we would expect to find higher SIDS rates in countries where bed sharing is common.

But just because bed sharing occurs frequently in cultures with low rates of SIDS, this does not mean that bed sharing *as practiced currently in Western cultures* is safe. Perhaps if we mimicked the sleep practices of cultures with low SIDS rates, this would be a good starting place for safe bed sharing. I asked Rachel Moon, lead author of the AAP's statement on safe sleep, if she could imagine a safe bed sharing environment, and she responded: "A pallet on the floor."[47] At first I laughed a little. "I'm serious," she said sternly. "In other countries, where they sleep with their babies and quite safely, that's what they do. In Japan, the futons are not like the futons they sell in the U.S. Here in the U.S., they're kind of foamy and cushiony, and in Japan, they're flat, and they're hard as rock. And that's what they sleep on." This may sound extreme, but if we believe bed sharing to be safe because it is found to be safe in other cultures, then we should be prepared to adopt as much of those sleep environments as possible.

Beyond the cross-cultural data, there's more evidence that sleeping close together might protect babies. The same European case-control studies that showed bed sharing was a risk factor for young infants also showed that putting a baby to sleep alone in another room, separate from his parents, roughly doubles the risk of SIDS. In one study, the authors estimated that 36% of SIDS deaths could have been prevented if babies had slept in the same room as their parents. In fact, sleeping in a separate room turned out to be a more important risk factor than bed sharing, which accounted for 16% of deaths. Add just those two risks together—sleeping too far and sleeping too close—and more than half of SIDS deaths could be prevented, according to the authors' estimates.[48] Other studies have found similar results.[49]

Nobody knows why room sharing is protective, but researchers speculate that it allows some sensory exchange between caregiver and baby. Babies are more likely to wake, and parents are more likely to check on them. Importantly, studies find that sharing a room with other children doesn't protect babies from SIDS. The presence of an adult devoted to caring for the baby seems to be a critical component of the protection of sleeping close. Studies haven't looked closely at how long the protective effect of room sharing is in play, but most experts assume that it is most important during the first six months of life, when the risk of SIDS is the highest.[50]

THE PHYSIOLOGY OF SLEEPING CLOSE

Anthropologist James McKenna is one of the most vocal advocates of the benefits of bed sharing. He began his career studying the social behavior of monkeys and apes, but when his son was born in 1978, he turned his attention to understanding the sleep of human infants and their mothers.[51]

In the 1980s, McKenna began detailed sleep physiology studies of breastfeeding mothers and babies at the University of California, Irvine.

Until McKenna's work, Western scientists had observed infant sleep patterns only in the context of solitary sleep. As an anthropologist, McKenna recognized that this was not the way mothers and babies had evolved, and nor was it the norm around the world. And as a father, McKenna was probably aware that it was not uncommon for American parents to bring their babies into bed with them.

In McKenna's studies, both the mothers and their babies were outfitted with a series of electrodes, belts, and thermocouples to take physiological measurements while they slept, including heart and respiration rates, brain activity, eye movements, muscle tone, and body temperature. Throughout the night, they were also videotaped by infrared cameras. McKenna and his colleagues recruited new moms, including some who routinely bed-shared at home and others who routinely slept separately. Each dyad spent three consecutive nights at the sleep lab, the first to allow the participants to adapt to the lab conditions, and the second and third to test, in random order, the effects of sleeping both together and apart.

McKenna and his colleagues found that when infants slept next to their mothers, they spent a little less total time in deep sleep than they did when they slept alone in a crib. They also had more short arousals from deep sleep but fewer arousals during active, or REM, sleep.[52] Like their infants, bed-sharing moms also woke more often, usually in response to their babies, but these arousals were short and didn't reduce their total sleep time.[53] Bed-sharing dyads also breastfed more during the night.[54] One of McKenna's findings that is rarely discussed is that bed-sharing infants had more episodes of central apnea (a period of more than three seconds without breathing or effort to breathe).[55] We can only speculate whether these differences in sleep are good or bad for babies, but what is clear is that whether babies sleep alone or with their mothers does affect their physiology.

McKenna contends that human infants evolved to sleep with their breastfeeding mothers and believes that these observed physiological differences are adaptive. He thinks that the more frequent arousals observed in bed-sharing babies may protect against SIDS, preventing babies from sleeping too deeply for too long and giving them practice in regulating their breathing during the night.[56]

McKenna's work and theories flipped the conventional Western thinking about infant sleep on its head, and many parents embraced his ideas. Breastfeeding moms had been sleeping with their babies, though perhaps quietly, and now there was science to tell them that maybe what felt right was also right for babies. Undoubtedly, many sleep and SIDS researchers felt challenged. Sleep researchers had been working on the assumption that when babies slept through the night, that was a good thing, and here was a guy telling them that frequent wakings were good for babies. SIDS researchers were finding an association between bed sharing and SIDS, and McKenna was telling them that maybe they had it all wrong.

What McKenna has taught us is that we don't really understand infant sleep if we study only infants sleeping in cribs. But there is still so much that we don't know. SIDS does seem to be related to a failure of babies to arouse from sleep, and it makes sense that if bed-sharing babies spend less time sleeping deeply and arouse more frequently, this might protect them. It makes sense that if an attuned mom is waking and briefly checking on her baby frequently during the night, this could decrease the chances that the baby will die in his sleep. But we really don't know if any of this is true. And unfortunately, the numbers still tell a different story. Although sleeping in the same room—close, but not on the same surface—has repeatedly been shown to be protective, among the many case-control studies of SIDS, only one has so far found bed sharing to be protective (and only in babies older than 3 months).[57] It's possible that there are some elements of bed sharing that are truly protective, like lighter sleep and more chances for a mom to check on her baby. And it may be that other elements of bed sharing, like the risk of the baby's face being covered with bedding, are hazardous.

If our foremothers slept with their babies from the dawn of our species, they probably also experienced both benefits and risks from the practice. Of course, the benefits of bed sharing in a cold cave would have far outweighed the risks, but for us, the risk-benefit calculation is a bit different. Anthropologists often say that human babies were "designed" to sleep with their mothers. But babies weren't designed; they evolved. If they evolved sleeping with their mothers, then this was probably an adaptive practice within the circumstances of their lives. It doesn't mean that it was or is a risk-free practice or that it is necessarily the best way for our babies to sleep in modern times. Babies didn't evolve sleeping in cribs, but most of our current data indicate that a crib next to the parents' bed is the safest place for babies to sleep.

McKenna's hypothesis is a fascinating one, and it's such an attractive idea, calling to mind the picture of a peaceful mother and baby sleeping together, hearts beating in synchrony. But it is still just a hypothesis, one we should study further, while keeping it in context with real life in the twenty-first century.

BED SHARING AND BREASTFEEDING

One of the most commonly cited benefits of bed sharing is that it facilitates breastfeeding. Human milk is rapidly digested, so breastfed infants need to feed frequently.[58] This is especially important during the first few months, and this need continues through the night as well as during the day.

In general, moms who bed-share are more likely to breastfeed for longer durations.[59] A U.K. study found that babies who bed-shared throughout the first year had five times the odds of still breastfeeding at 12 months compared with those who slept alone.[60] However, this doesn't mean that bed sharing necessarily causes more breastfeeding; it's just a correlation. Maybe bed sharing makes breastfeeding easier, or maybe breastfeeding makes bed sharing a more attractive choice. Or perhaps both choices are part of a larger philosophy about raising babies. What these studies tell us is that bed sharing is common among breastfeeding moms, and it seems likely that the two practices mutually reinforce each other. If you're breastfeeding a baby every couple of hours, keeping that baby right next to you might make nighttime easier. Open your shirt, pull baby in closer, and voila—you're feeding.

Careful bed sharing may carry a small risk with young infants, but sleeping with babies on couches or armchairs is very dangerous. There's some evidence that the percentage of SIDS deaths involving cosleeping on couches is increasing.[61] In a U.S. survey of more than 6,000 moms, most of them breastfeeding, 55% said they fed their babies at night on a chair or sofa. Of these, 44% said they sometimes fell asleep while feeding.[62] If you're feeding your baby during the night and you think there's a chance that you'll doze off, consider the safety of the environment around you. You'll be safer and more comfortable if you feed in a bed that you've set up for safe bed sharing, just in case, than trying to prop yourself up on a couch and hoping for the best. And studies show that when you bring your baby into your bed for feeding and comfort and then return him to his own bed for sleep, there is no added SIDS risk.[63]

Breastfeeding also appears to reduce the risk of SIDS by about half.[64]

⁎ SAFE SLEEP RECOMMENDATIONS

- Always put your baby down for sleep on his back.
- The safest place for your baby to sleep is in a crib or bassinet in your room. Room sharing is recommended for about the first six months of life.
- Avoid sleeping with your baby on a couch or chair, or feeding in this position if you think you might fall asleep, because this dramatically increases the risk of SIDS and suffocation.
- Bed sharing definitely increases the risk of SIDS if the baby was born prematurely or was exposed to cigarette smoke (before or after birth), or if mom or other sleep partners have been drinking alcohol or taking drugs or medications that might impair responsiveness.
- Bed sharing during the first three to four months of life is thought to increase the risk of SIDS, though some researchers disagree. Wherever your baby sleeps, know that SIDS risk is greatest during this time, so use extra care.
- Your baby should always sleep on a firm and flat mattress.
- Safety considerations for cribs: Ensure that the crib meets current safety standards. Do not put bumper pads, bedding, pillows, stuffed animals, or toys in the crib. A tightly fitted sheet is all that is needed.

This may be because breastfed babies arouse more easily from sleep than those fed formula,[65] or perhaps it's because breastfed infants tend to have fewer respiratory infections, which can increase SIDS risk.[66] This is a factor to add to your personal risk-benefit analysis for bed sharing; if bed sharing means you're more likely to continue breastfeeding, then the protection of breastfeeding may offset the risk of bed sharing.

RECONCILING THE DEBATE IN THE REAL WORLD

When I began the research for this chapter, I hoped that, with enough reading and thought, I could eventually arrive at a clear understanding of safe infant sleep. As is probably obvious by now, I haven't found One Right Answer to this question.

⁎ Safety considerations for the bed sharing environment: Remove mattress from the bed frame and place it on the floor and away from walls. Side rails are not recommended, since they can create a space for entrapment. Minimize the use of pillows and blankets, and keep them well away from the baby. Never leave a baby to sleep alone in an adult bed, and avoid bed sharing with other children, nonparent adults, and pets.

⁎ Keep your baby warm using clothing and a sleep sack rather than loose bedding. Don't overbundle him; dress him just enough to stay warm. The head is an important area for heat dissipation, so avoid using a hat for sleep. Ideal room temperature is thought to be 61° to 68° F.

⁎ Giving your baby a pacifier for sleep (even if it falls out after a few minutes) may help prevent SIDS.

P. Blair and S. Inch, "The Health Professional's Guide to: 'Caring for Your Baby at Night,'" UNICEF U.K. Baby Friendly Initiative, 2011, https://www.unicef.org.uk; AAP, "Policy Statement: SIDS and Other Sleep-Related Infant Deaths: Expansion of Recommendations for a Safe Infant Sleeping Environment," Pediatrics 128 (2011): 1030-9.

In my quest for answers, I spent some time talking with Dr. Rachel Moon, the chairperson of the AAP's Task Force on SIDS and the lead author of the 2011 policy statement recommending against bed sharing.[67] As a practicing pediatrician and mother of two exclusively breastfed babies, Dr. Moon is well aware of the challenges of infant sleep. Still, she sees the question of bed sharing as a fairly simple one: in her view, it just isn't safe. "Nine times out of ten, it [bed sharing] will be okay," she told me. "But I come from a different perspective because I spend my life talking to people where things *haven't* gone okay . . . If you've seen one of those cases, you just don't want to take the chance. Why take the chance?"[68]

Moon carries with her the weight of grieving families. She knows as well as anyone that the data aren't perfect, but in her mind, bed sharing just isn't worth the risk. The worst-case scenario is just too grave. A bed sharing advocate like Jim McKenna, on the other hand, spends his time working with families for whom sleeping with their babies is a valued practice. Both Moon and McKenna have the best of intentions in the way they approach this topic, but both have unavoidable emotional biases. The difference is that McKenna is telling us what many of us want to hear: that it is natural to keep our babies close and that they're safe there. Moon's message is far less romantic and reassuring. From a public relations standpoint, I think that Moon has the more challenging job.

The truth is that no study design can fully describe something as complicated as how a mother and baby sleep. Scientists and parents can argue about the safety of bed sharing all they want, but we may never know who is right. And to be honest, I'm not sure it really matters. If it were only about safety, I believe that most parents would take the AAP's advice to heart and avoid bed sharing, just as most have adopted the Back-to-Sleep recommendation. But bed sharing is a more complex behavior. Parents have been sleeping with their babies forever, and there are cultural, biological, and physiological reasons why we may be drawn to hold our babies close at night. Parents will continue to bed-share, with or without the approval of the AAP.

Since I can't provide you with a single recommendation on this complex topic, I'll leave you with this advice: whatever route you choose, do your best to make it safe. You understand the controversy, and you have the ability to make your own risk-benefit analysis. Science can never quantify all of the risks and benefits of bed sharing or, for that matter, any parenting practice. It can give us some clues at the population level, but you have to factor in how you and your baby feel about the matter as well. Given all the random and wonderful human variation among parents and babies, it would be unreasonable to think that all of us should do everything the same way.

7 ⁂ IN SEARCH OF A GOOD NIGHT'S SLEEP

(Or Something Like It)

I have a vivid memory of a conversation with my mother in a dark hallway during Cee's first week of life. It was some wretched hour in the middle of the night. Cee was feeding every 60 to 90 minutes, and each time she woke, the labor-intensive routine of breastfeeding (when we were both still getting the hang of it) and helping her settle back to sleep left little time for either of us to sleep between feedings. My husband, meanwhile, was remarkably good at sleeping through most of this.

My mom was staying with us to help for a few weeks, and she must have woken with Cee's cries that night. She peeked into my bedroom. "Is everything going okay, Alice?" she asked.

I felt the back of my throat tighten and surprise tears well up in my eyes. In fact, everything was not going okay. I was exhausted, my nipples were sore, and I felt that surely it wasn't supposed to be this hard. Weren't newborns supposed to sleep all of the time? Was there something wrong with my baby? Or with me? Or my milk? As we talked in the hallway, my mom responded in what I now know was the most perfect way she could have answered my concerns. She didn't offer a solution or even reassurance that it would get better anytime soon. She simply said: "This is completely normal. This is how newborns sleep and feed."

Somehow, knowing that Cee's sleep was normal made it bearable. Of course, right around the time I started to understand newborn sleep, her sleep pattern changed. I had so much to learn, and this was just the beginning. Cee is 3 years old now, and it's still sometimes a struggle to help her—and us—get enough sleep.

Having a baby changes everything about sleep, and nothing can really prepare us for this. I coped with our sleep struggles by learning as much as I could about the science of infant sleep. I found a wealth of fascinating research that helped me to understand what normal infant sleep looks like, how it develops, and how our parenting shapes our babies' sleep development. I put that knowledge into action to help my family find a good night's sleep, and I hope this chapter can help you do the same.

SLEEP IN EARLY INFANCY: WHAT IS NORMAL?

Newborn babies sleep as much as 16 to 18 hours of the day, but they don't always sleep when we hope they will.[1] Instead, they sleep just about as much during the day as they do at night. Their sleep is also fragmented into short cycles, and they often wake up hungry.

Breastfed babies, in particular, need to feed frequently. Human milk is more easily digested than formula, so breastfed babies need their small tummies refilled often, both night and day.[2] Frequent nursing also helps to establish mom's milk supply. In one study, breastfeeding mothers were encouraged to feed their babies frequently during the first weeks of life, and at 15 days old, their babies were drinking more milk and had gained more weight than babies fed on a three- to four-hour schedule.[3] All newborns, whether breastfed or formula-fed, should be fed on demand, day and night.

Newborn babies cycle between active (rapid eye movement, or REM) sleep and quiet (non-REM) sleep, their sleep split about equally between the two types. They usually fall asleep in active sleep, when you'll see their eyes moving beneath closed lids, animated expressions crossing their faces, and twitches and movements in their limbs. They also wake easily from active sleep, so if your baby falls asleep in your arms, trying to move her during this stage might be the end of the nap.[4]

After about 20 to 25 minutes of active sleep, newborns shift into quiet sleep. Their breathing becomes slow and rhythmic, and their bodies are still. In this deeper sleep, you can usually transfer a baby from your arms without waking her. She'll be in this phase for about another 20 to 25 minutes, after which she may continue to another active-quiet sleep cycle.

However, making the transition from one cycle to the next is tricky for newborn babies. They'll often wake at this time and either be truly awake or need help getting back to sleep to begin another cycle.[5]

Thankfully, sleep patterns mature rapidly in the first few months of life. By the second month, babies get better at transitioning from one cycle to the next without waking, and they often have a long sleep bout during the first part of the night. They'll begin to start their sleep in quiet sleep and spend more time there. Active sleep will decrease and shift toward the early morning hours, which may mean your baby is more restless during the second part of the night.[6] As their stomachs grow, they can go longer between feeds, allowing their sleep to consolidate into longer periods.[7]

During these first few months, babies gradually start sleeping more during the night and less during the day. They're adapting to the outside world, where nights are dark and quiet and days are bright and busy. These light and dark signals help babies develop a circadian rhythm, synchronizing them with the 24-hour day-night cycle. When daytime light hits the circadian control center of the brain, it coordinates wakefulness throughout the body, including an increase in body temperature and cortisol and suppression of the sleep hormone, melatonin. When nighttime brings darkness, melatonin is released, allowing the body to rest and sleep.[8]

In the 1990s, psychologist Kate McGraw was working in the Sleep Study Unit of the University of Texas Southwestern Medical Center. She was so interested in the development of infant circadian rhythms that she enlisted her own newborn as a study subject. She and her husband observed their son, Tyler, day in and day out for the first six months of his life. They took his temperature hourly and collected his saliva weekly to measure his melatonin production. They exposed him only to natural sunlight, using no artificial lighting during the night, and they kept their whole family on a predictable schedule of sleeping and eating. Tyler, on the other hand, was allowed to breastfeed whenever he was hungry and sleep when he was tired. Kate watched as baby Tyler began to show signs of a circadian rhythm, and she and her colleagues reported their findings in a paper published in 1999 in the journal *Sleep*.[9]

Just one week after his birth, Tyler's body temperature was dropping to its lowest point during the night, just as happens in adults. A week or so later, he was sleeping more at night and being more wakeful during the day. Consistent wake periods of 90 to 120 minutes emerged in the morning and before bedtime in the evening; these were peak play times for Tyler. By day 45, he showed a strong melatonin rhythm, high during the night and low during the day. By the end of his second month, he was basically sleeping from sundown to sunrise, with a few brief awakenings during the night for feedings.[10]

This study was fascinating because of its detail, but Tyler was just one infant among many. Other studies, though, have shown similar timing for the emergence of a circadian rhythm, although the careful control of Tyler's natural and social environment may have helped him find his rhythm a bit earlier than average.[11] There's also a lot of inherent variation among babies, but most are sleeping much more during the night than during the day by the time they are 2 or 3 months old.[12]

HELPING YOUR BABY LEARN NIGHT FROM DAY

During your baby's first few weeks, sleep feels very disorganized. Some naps will be long and some short. Your baby may be sleepy during the day and wakeful at night. Your goal during this time is to help her rest when she's tired and to feed her when she's hungry. It's a simple but exhausting job. The good news is that as you show your baby the natural signals of night and day, she'll quickly begin to develop a circadian rhythm and more sustainable sleep patterns.

At night, keep the lights as dim as possible and the environment quiet and relaxed, even if you're feeding frequently. Breastfed babies also have the benefit of melatonin in breast milk, which is high during the night and undetectable during the day. Since it can take babies a month or two to develop their own melatonin rhythm, milk melatonin can help bridge this gap to promote nighttime sleep.[13]

During the day, let your baby be part of the activity of the house. Even as she sleeps, keep her in a light room. One study found that among 6- to 12-week-old babies, those who were exposed to more light between noon and 4:00 p.m. slept better at night.[14] After the first few months, most babies nap better in a quiet and dark place. You will want to respect that, but by then your baby should—hopefully—have the day and night straightened out.

Giving your baby these natural signals of light and activity allows her to develop a circadian rhythm that matches your own. Illustrating the

✲ SUPPORTING HEALTHY NEWBORN SLEEP

- Help your baby learn the rhythm of days and nights. Let your home be bright, with normal activity, during the day. At night, keep it dark and quiet, with lights and monitor or TV screens very dim or off.
- Understand newborn sleep cycles. When your baby is in active sleep, she may wake easily, and she may be noisy. Don't assume that every grunt and sigh represents a wakeful, hungry baby. Before intervening, watch to see whether she's just moving in her sleep, or is waking briefly, or is really asking for your help.
- Observe your baby to learn how long she can be awake before she gets tired. Help her get comfortable for sleep before she's overtired.
- Develop a simple, soothing bedtime routine.
- Keep a log of your baby's sleeping and feeding so that you can see patterns emerge and have a little predictability for your days and nights. There are smartphone apps that make it easy to do this.

importance of environmental and social cues, one study found that both preemie and full-term babies develop strong day-night rhythms within 9 to 10 weeks of coming home from the hospital, despite the relative neurological immaturity of the preemies.[15] One baby in this study still hadn't developed a circadian rhythm at 5 months of age and was just as wakeful during the night as during the day. It turned out that when this night owl woke during the night, he was fed in a bright room; he was getting confusing signals about day and night.

Despite all of the studies on infant sleep, little work has been done on wakefulness. But from a practical standpoint, an awareness of how long your baby can handle being awake is important to timing good sleep. Baby Tyler, whose sleep and wake cycles were tracked by his scientist mom, showed distinct periods of 90 to 120 minutes of wakefulness in the morning and in the late afternoon by the time he was 1 month old.[16] As wakeful periods emerge in your baby, those times of day will be best for playing and interacting. And as the wakeful period winds down, keeping an eye on the clock and your baby's signs of tiredness will allow you to prepare her for sleep before she becomes overtired. The amount of time that your baby can comfortably be awake will increase as she grows; one study found that, on average, the longest period of wakefulness was about 2 hours at 3 months of age and 3.5 hours at 6 months.[17]

It's also never too early to develop a soothing bedtime routine, a sequence of calming rituals that you use each night to get your baby ready for sleep. Use a similar but simplified routine for naps. A randomized controlled trial showed that a routine of bath, massage, and snuggles helped older infants and toddlers to fall asleep faster, sleep more during the night (an average of 36 more minutes!), and wake in a better mood in the morning.[18] What you include in the routine is probably not important, so long as it is soothing and predictable, with the same sequence each night.

SLEEP IN THE FIRST YEAR: WHAT IS NORMAL?

Once your baby can tell her nights from her days, sleep should get a lot easier. The most rapid sleep consolidation happens within the first several months of life, and by 3 to 4 months of age, many babies are sleeping for nice 8- to 10-hour chunks without waking their parents.[19] However, the fact that many babies do this doesn't mean that your baby will or even should. When it comes to infant sleep, there is a wide range of normal.

When I polled my blog readers about when their babies began sleeping for eight-hour stretches, the responses ranged from as early as 5 weeks to well into the toddler years. It's not a scientific sample, but it gives you an idea of how pointless it is to worry about what is normal. If your baby sleeps through at 2 months, she's normal; lots of babies catch on early. But then again, if she doesn't sleep through the night until 2 years, she's normal, too. Although there are things you can do to help your baby sleep through the night sooner (which I discuss later in this chapter), there's also a lot of inherent variability from child to child. Between 6 months and 4 years of age, genetics explains about half of the differences in sleep patterns from one child to the next.[20]

It is also completely normal for a baby to be sleeping through the night for some time and then to start waking again in later infancy, just when you think you have your sleep back.[21] These sleep setbacks can occur because of teething, illness, developmental leaps like learning to crawl (and wanting to practice during the night), and separation anxiety, among other things. Many babies are still waking their parents at least once each night at 12 months.[22] In fact, half of toddlers (12 to 35 months) and one-third of preschoolers (3- to 5-year-olds) in the United States wake their parents at least once on a typical night.[23]

Breastfed babies usually take longer to sleep through the night, a factor that has not always been considered in studies of "normal" infant sleep patterns. When anthropologist Helen Ball surveyed British mothers about their babies' sleep habits, she found that two out of three babies were no longer waking their parents for a feed by 3 months of age.[24] However, 96% of these dreamers were formula-fed. Sleeping through the night at this age was rare for a breastfed baby. Among the babies still waking at 3 months old, about half were formula-fed and half were breastfed. Formula feeding *may* help with sleep consolidation, but it isn't a golden ticket by any means.

HOW BED SHARING AFFECTS SLEEP

Many parents choose to bed-share with their babies, particularly if the mom is breastfeeding. Bed sharing often isn't the original plan, but parents soon discover that physical closeness at night makes it easier to monitor, feed, and soothe the baby, and the carefully decorated nursery ends up abandoned for many months.[25] Other parents find that everyone sleeps better when the baby is in her own bed. It's a personal choice, one that should include an evaluation of safety concerns (discussed in chapter 6), sleep quality of the baby and parents, and family dynamics.

Most studies show that bed-sharing babies—and their mothers—tend to wake more often than solo sleepers.[26] However, some studies find that total nighttime sleep does not differ, suggesting that the bed-sharing babies wake more often but for shorter periods of time.[27] Others find that bed-sharing babies get a little less total sleep than those who sleep alone.[28] Despite a more fragmented (and maybe shorter) night of sleep, many families choose to bed-share because it helps to get them *more* sleep overall, particularly in the early weeks and months of the baby's life and especially if mom is breastfeeding.[29] In breastfeeding moms and babies, sleep cycles can become synchronized so that waking is less disruptive,[30] and some moms find that they can doze restfully while their babies nurse. Many families that choose to bed-share accept that night wakings are normal and to be expected.[31] Others don't adapt well to frequent wakings and find serving milk all night to be exhausting. Just as many parents swear that bed sharing is the key to sleep, many others choose not to bed-share because they say it interferes with good sleep.[32]

The link between bed sharing, breastfeeding, and night waking was nicely demonstrated in a 1986 study published by Harvard child psychologist Marjorie Elias and her colleagues.[33] They studied 32 breastfeeding moms in suburban Boston. Half of the mothers were recruited from the local La Leche League (LLL) group and the other half were considered the "standard care" group. Although both groups were overwhelmingly white, middle-class, and well educated, their parenting practices were quite different. Standard care moms nursed their babies about five to seven times per day during the first year, and most weaned between 7 and 20 months. LLL moms breastfed about 11 times per day during the first year, and all but one were still breastfeeding at 24 months. About 25% of the standard care moms bed-shared for at least part of the night during the first year, while bed sharing was as high as 88% in the LLL group. In other words, the LLL moms were practicing a more intensive style of parenting, similar to today's attachment parents.

These different parenting philosophies and practices were reflected in the babies' sleep patterns. Compared with the LLL babies, the standard care babies had longer sleep bouts during the night and, beginning at 10 months, had more total sleep per 24-hour period. The researchers tried to tease apart the effects of breastfeeding and bed sharing on sleep patterns, but they found that both were important. When they looked at the longest sleep period (without waking a parent) of 2-year-olds, they found that those who both breastfed and bed-shared slept for an average of 4.8 hours at a time, those who breastfed but slept alone slept for about 6.9 hours, and those who were weaned and slept in their own beds slept for 9.5 hours.[34] Based on this study, if your breastfed and bed-sharing baby is waking frequently, it probably isn't because she's incapable of sleeping through the night but because the cozy bed with her favorite person and favorite food isn't exactly encouraging it.

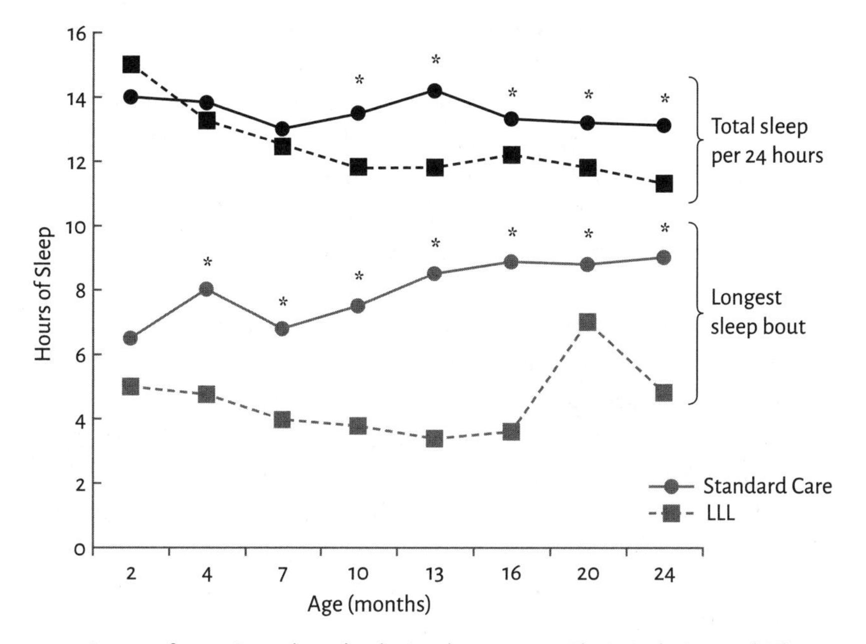

Impact of parenting style on developing sleep patterns. The La Leche League (LLL) babies were breastfed more frequently and for a longer duration in childhood, and most were bed-sharing. The standard care babies were breastfed less frequently, weaned earlier, and less likely to bed-share. Across infancy, LLL babies had shorter sleep bouts, and in late infancy, shorter total sleep per day. Asterisks indicate statistically significant differences between groups at a time point ($P < 0.05$). *Source:* M. F. Elias et al., "Sleep/Wake Patterns of Breast-Fed Infants in the First 2 Years of Life," *Pediatrics* 77 (1986): 322-29

All of this raises the question, is one or the other pattern—the frequent feeding and waking associated with bed sharing or the more consolidated sleep found in babies that sleep alone—better for babies? We really don't have much evidence either way. When breastfed, solo-sleeping babies have more consolidated sleep, they tend to make up for fewer night feedings by indulging in a big morning feed, and there's no apparent difference in their growth rates.[35] Although bed sharing is often promoted as a way to build a stronger attachment relationship, what little research we have on this idea shows that this isn't necessarily the case. Wherever your baby sleeps, what seems to be most important to attachment *and* good sleep is emotional availability—being sensitive to your baby's cues and responding appropriately.[36] In this sense, the best sleeping arrangement is probably the one that feels right for you and works well for your baby.

In many families, bed sharing works well in early infancy but loses its sweetness as the baby approaches toddlerhood. I talked about this with a local pediatrician. She values bed sharing herself; her children were born while she was in medical school and residency, and they slept in her bed to make up for their time apart during the day and to help maintain her breast milk supply. But, she told me, she sees moms in her clinic every week who are breastfeeding and bed sharing with 12- to 18-month-olds and are just exhausted. "These babies wake frequently, especially in the second half of their sleep. They are used to nursing to sleep, and they need that every time they wake during the night."[37] Some parents ride out this period, knowing that their children will eventually outgrow it. Others "night wean," continuing to bed-share but explaining to their toddlers that the milk bar is closed for the night. Still others decide it is time for their baby to have her own bed, and the frequent wakings will often resolve on their own.

NIGHT WAKING AND SELF-SOOTHING

In the late 1960s, a young psychiatrist named Thomas Anders was studying infant sleep patterns in the newborn nursery at Montefiore Hospital in New York City. Another researcher was working under the same mentor, only she was studying the development of sleep in kittens. As Anders told me, a romance developed between the two, and they were married. Their first child, Michael, was born in 1972.[38]

In his research on the sleep of newborn babies, Anders had been using a technique called polysomnography, a set of physiological measurements that required attaching an array of harmless electrodes to a baby's body. In those days, newborn babies spent their nights in the hospital nursery, and most moms agreed to let Anders conduct these studies as their babies slept. However, he had a hard time getting moms to bring their babies back into the sleep lab for longitudinal studies, to see how the infants' sleep developed over the first months of life. Parenting is hard enough in those early days, and the last thing those moms wanted to do was bring their little babies into a hospital sleep lab and put them to bed with a bunch of electrodes. Even for the sake of science.

So when his first child was born, Anders thought: "Here's the perfect subject for a longitudinal study." But, he told me, "When it came time to put the electrodes on him and do the polygraphic recordings, both my wife and I sort of looked at each other." They knew that the recordings wouldn't hurt their little boy. But the electrodes had to be attached with a type of glue, which would get stuck in his hair, and the whole thing was probably irritating to a baby. Anders and his wife couldn't quite bring themselves to make their son a guinea pig in this way.[39]

What I love about this story is that the experience of becoming a parent pushed Anders to find a new and better way to answer his research questions. He ditched polysomnography and instead began using what was then a brand-new technology: time-lapse video. He could skip all the electrodes and the glue and wires. All that was needed was a video camera mounted above a baby's crib at home. Watching these videos of babies sleeping, Anders could easily differentiate active sleep from quiet sleep.[40] At the time, he was interested in how sleep architecture changed with age, thinking that this might be a window into how a baby's brain was developing.

As he watched these videotapes, tracking cycles of active sleep, quiet sleep, active sleep, quiet sleep, and so on, Anders noticed something totally new and unexpected. He noticed that every single baby woke up at least a few times during the night, even the ones who had appeared to be "sleeping through the night."[41] Before this discovery, sleep researchers assumed that babies who slept through the night really did just that—slept the entire time. Instead, Anders found that all babies woke up at some point during the night. Their longest sleep bouts lengthened during the first year, but it was unusual for any baby to sleep for more than six hours at a time.[42] As Anders described it: "It is normal, it is physiologic, and it is expected that babies even at a year wake. I would think that these brief awakenings are necessary for the nervous system."[43]

All the babies in Anders's studies woke during the night, but what differed was how they responded to an awakening. Some babies cried when they woke, and Anders labeled these babies "signalers." Others woke, looked around, perhaps grabbing a teddy bear or sucking their thumbs, and then went back to sleep. Anders called these babies, the ones who didn't alert their parents to their awakenings, "self-soothers."[44]

Now, this discovery may not sound so groundbreaking to you. You might have already read about "self-soothing" in one of the baby sleep books on your nightstand. But at the time, this was a really important finding. Anders had essentially defined, in an objective way, what generations of parents meant when they said their baby was a "good sleeper": a baby that woke in the night but was able to go back to sleep on her own. And with this discovery, Anders and others could begin to identify factors that were different between self-soothers and signalers.

WHAT MAKES A SELF-SOOTHER?

Overwhelmingly, and across cultures, babies that self-soothe usually have one thing in common: they are put in their beds at night while they are awake.[45] They aren't fed or rocked or bounced to sleep. Instead, they manage the transition from wakefulness to sleep on their own at the beginning of the night, and having this skill, they're able to do the same thing in the middle of the night as well. It's as if they wake up, look around, and think, "Oh yeah, here I am in my crib still. Nothing interesting going on here. Yawn. Guess I'll go back to sleep." Signaling babies, on the other hand, are often actively soothed to sleep at bedtime and then gently set down in their cribs. When these babies wake up, they cry. Perhaps they are thinking, "Wait a second! Last thing I knew, I was nuzzled against my mama, drinking my favorite drink, listening to her sing my favorite song. Now I'm lying here alone in this crib. Mama? Mama! Bring back the milk and the warmth and the music!" In other words, the baby that falls asleep with help at the beginning of the night is likely to call out for help in the middle of the night. (As with everything in parenting, there are certainly individual exceptions to this rule, but in studies that look at this question in a group of babies, this pattern almost always emerges.)

Anders explained to me that "falling asleep is a *learned* phenomenon" and that, according to his data, most self-soothing babies learn to fall asleep on their own by about 4 to 5 months of age. And because self-soothing is a learned skill, it takes a little practice.

Your baby gets to practice self-soothing to sleep when you start putting her down in her bed while she's still awake. This may mean ending her feeding session before she falls asleep completely or even waking her up a bit as you put her to bed. It might mean avoiding previous "active" soothing methods, including rocking, bouncing, and walking to sleep. The goal is to let your baby try falling asleep on her own. Anders told me that at this stage, if the baby protests being put down to fall asleep, "You can quietly, soothingly, sit by the baby and talk to the baby, pat the baby . . . You can help the baby to not be so upset and fall asleep."[46] This might be the beginning of a gradual learning process, and you can offer your support without jumping in to take over for the baby.

One study from Anders's lab showed that at 12 months of age, babies were more likely to be self-soothers if their parents waited just a few minutes before responding to their night wakings when they were 3 months of age.[47] These parents weren't neglecting their babies; average wait time during this study was just three minutes. Maybe that was just long enough to give the babies a chance to try soothing themselves, maybe sucking a thumb or fingers, stroking a soft toy, finding a pacifier, and so on. And most babies do make some attempts to self-soothe after waking during the night, even if for just a few seconds and regardless of whether or not they end up crying for help.[48] Many of the babies who were usually signalers had occasional awakenings during which they quietly soothed themselves back to sleep without waking their parents. Most babies have some ability to self-soothe, and this can develop further over time with some brief opportunities to practice.

A little wait time also recognizes that when babies are in active sleep, they can be incredibly noisy and active. They might grunt, whimper, cry (briefly), twitch, flail their limbs around, and even open their eyes, all without truly waking. A parent who comes running at every little peep might end up waking a sleeping baby, and if that happens enough times, waking can become a habit. With this in mind, if your baby wakes you during the night, you might take a few minutes to pause and listen before reacting. If she's fussing quietly, she may not really need your help. But if she sounds distressed, then there's not much point in letting her get more worked up.

A few randomized controlled studies have tested parent education programs that teach strategies to help infants learn to self-soothe, including putting the baby down awake and waiting a moment before responding after an apparent night waking.[49] When parents received this advice, their babies slept longer and woke their parents less often during the night by the time they were a few months old. Because these were prospective and randomized studies, they clearly showed that parents can shape their babies' sleep and that giving brief opportunities to self-soothe from a young age can lead to more restful sleep.

* TIPS FOR SUPPORTING SELF-SOOTHING

- Put your baby to bed when she is sleepy but awake.
- When your baby wakes during the night, wait a few minutes before responding, especially if she's just making quiet noises.
- Offer your baby a special lovey (something small that won't pose a choking or strangulation hazard). It may help her feel more secure at night. (Or it may not, but it's worth a try!)

OF BLANKIES, LOVIES, AND PACIES: SLEEP AIDS AND SELF-SOOTHING

Tom Anders also found that self-soothing babies often use some kind of sleep aid, like a pacifier, doll, or some other kind of "lovey."[50] After making this discovery, he wondered whether he could teach *all* babies to self-soothe by giving them just the right lovey. To test this hypothesis, he and his colleagues asked 30 moms to infuse a large T-shirt with their body odor by breastfeeding in it for a few nights.[51] Then they knotted up each mom's shirt and put it in her baby's crib. As controls, 30 babies received a washed shirt without mom's scent. The expectation was that the babies snuggling with their mama's scented shirt would be soothed by that familiar smell, helping them to bridge the night wakings without needing their mother.

Well, the result of the T-shirt experiment was not as expected. Some of the babies snuggled with the smelly T-shirts, and some of the babies snuggled with the clean ones. Others used pacifiers, sucked their thumbs, or adopted other lovies, and some didn't adopt a sleep aid at all. Regardless, the presence of mom's smell didn't teach the babies to self-soothe. And although babies who self-soothe often do use sleep aids, the choice of a lovey is a very personal thing. Most babies want to choose their own.[52]

Although he may not have felt this way at the time, Anders is glad that the T-shirt lovies didn't work.[53] What the experiment showed was that self-soothing can't be forced on babies. You can give them opportunities to learn. You can place soft and smelly objects close to them or even gently show them how to find their thumbs. But you can't make a baby self-soothe. Every baby will respond differently to these opportunities to learn, and when they do learn to self-soothe, it will be in their own way.

Sleep aids are also a telling example of how parenting culture and philosophy affect infant sleep. In many cultures, they're virtually unheard of. Guatemalan and Korean babies, who usually fall asleep in physical contact with their mothers, rarely use sleep aids.[54] In a Cleveland study, children who bed-shared were less likely to use a sleep aid than those who fell asleep alone,[55] suggesting that when a parent actively helps a baby transition to sleep, the parent becomes the sleep aid. When a parent isn't there, babies may adopt some other comforting object to help them with that transition. Babies' use of inanimate sleep aids is expected in many Western cultures, and it probably goes hand-in-hand with independent sleep. But sleep aids and independent sleep are uncommon in many other cultures. Who are we to say that one or the other is better for children's development?

WHEN SELF-SOOTHING IS NOT THE GOAL

This is a good time to point out that much of what I'm saying here about desirable sleep for a baby, while rooted in the scientific literature, is still decidedly biased toward Western cultural beliefs and my own experiences. When Cee dropped her last middle-of-the-night feeding and started "sleeping" (or more accurately, self-soothing) through the night at around 8 months of age, I both celebrated the full night of sleep for myself and felt proud for her. To me, this was a developmental milestone that represented her growing autonomy and ability to self-regulate. There was also a part of me that mourned the loss of our quiet time together in the night, just as I mourned the passing of nearly every baby stage.

But not everyone shares my perception of self-soothing as a milestone or has the expectation that self-soothing is something with which babies should be concerned. For example, on the topic of how to respond to night wakings, attachment parenting advocates Bill and Martha Sears write: "If you get to your baby quickly before he completely wakes up, you may be able to settle him back to sleep with a quick laying on of hands, a cozy cuddle, or a warm nurse. If you parent your baby through this vulnerable period for night waking, you can often prevent him from waking up completely."[56] Based on what we know about self-soothing, this strategy might actually wake a sleeping baby or teach her to look for you upon waking rather than try to self-soothe. But lots of parents around the world approach nighttime parenting in this way, and their children will eventually self-soothe and sleep through the night, even if a few years down the line.

Understanding self-soothing also helps explain why breastfeeding and bed sharing may be associated with more night wakings. Several studies have shown that it is probably not breastfeeding per se that causes more wakings but rather breastfeeding *to sleep*, at least in older babies.[57] That's consistent with this same idea that if the baby falls asleep with the active involvement of a caregiver, she'll ask for the same kind of involvement after a normal night waking. Likewise, if a bed-sharing baby falls asleep snuggled against her mama, she will seek those same conditions upon waking in the night. And with bed sharing, it is much harder to wait a minute or two before reaching out to pat the baby or offering a breast. This means that the baby can grow accustomed to constant outside comfort and help with transitioning from one sleep cycle to the next.

If that help is always there and it doesn't interrupt the parents' sleep much, this system can work very well. This was demonstrated in a study comparing the sleep practices and problems of American and Japanese families.[58] In the Japanese families, most babies not only bed-shared but also fell asleep and remained in body contact with mom throughout the night. The mothers in these families reported few problems with bedtime struggles or night wakings. Some of the American families bed-shared too, but they were much more likely to report problems with it. One striking difference between the two cultures was that the American bed-sharing children were more likely to fall asleep outside the bed (maybe rocking to sleep, in a stroller, etc.). Since these falling-to-sleep conditions wouldn't be matched in the middle of the night, this might have caused more sleep disturbances, whereas the Japanese children could fall asleep knowing that they'd find things just as they left them if they woke in the night. If you're bed sharing, you might consider how you can match the going-to-sleep expectations with the middle-of-the-night expectations.

YOUR SLEEP PHILOSOPHY SHAPES YOUR CHILD'S SLEEP

In a 2009 study, Israeli researchers asked pregnant women to describe their philosophies about infant sleep, presenting them with 14 vignettes of baby sleep situations and asking them how they would respond.[59] For example: "Kevin is an 8-month-old boy. He is described as a very active and alert child . . . During the night, Kevin wakes up a number of times and has difficulty falling asleep." The women completed the questionnaire again several times during their babies' first year. None of the moms were extreme in their answers, but when the results were tallied, the researchers found that some moms put more emphasis on encouraging independent sleep and others were more likely to interpret night waking as infant distress and emphasize active parental soothing. They all turned out to be good, responsive moms; some were just more intense in their responsiveness than others.

The women who, while pregnant, expressed a sleep philosophy that assumed a waking baby was a distressed baby who needed help from her parents ended up being more involved in soothing their babies to sleep, and their babies woke more during the night at 6 months of age.[60] In a follow-up study, there were similar correlations between the mothers' sleep philosophies at 12 months and their children's sleep when they were 4-year-olds.[61] What this study shows is that your sleep philosophy helps to shape your child's sleep. If you believe your baby can self-soothe, and you let her try it, chances are good that she will self-soothe. If you think your child needs your help to sleep, then chances are that she will.

The authors of this study are quick to point out that "it would be wrong to conclude from these findings that parents should abstain from approaching their infants at night in order to facilitate good sleep patterns. Undoubtedly during the first months of life, infants need their parents for comfort and regulation, while gradually these functions shift from the caregiver to the infant. In the course of the infant's development, most parents sensitively balance between their infants' need for proximity and their need to develop separateness and autonomy. However, some parents find it difficult to keep this balance and adopt an unbalanced approach of either over-involvement or avoidance."[62]

My personal philosophy is that, as a parent, it is my job to find that balance of when my child is ready to try something on her own and when she needs help. I believe that babies are born with the desire and ability to learn and grow, and we should respect that, even as we wish we could freeze in time these moments of sweet baby bliss. Overnight video studies show that some babies are capable of self-soothing even as young as a few weeks of age.[63] Babies are extraordinary, aren't they?

We'll know what they are capable of only if we are willing to step back and give them a chance to try something on their own before we jump in to help. Think of other developmental milestones, like learning to sit up. If, every time we saw our babies working at getting upright, grunting with effort, furrowed brow, we jumped in and propped them up, how would they learn? I think of nighttime parenting strategies as a continuum of responsiveness, and we parents must try to find our own sweet spot, where our babies know that they are loved and supported but also have a chance to grow and develop their own sense of autonomy. But all of that said, I'd much rather err on the side of being a bit too responsive. And, this is just *my* philosophy. It's what I'm comfortable with, and it is how I make sense of the daunting task of parenting a child day and night. Your philosophy might be different.

Whatever your philosophy, it's important to remember that every baby is different. We may be able to help shape our babies' sleep, but we can't control it. Some babies just naturally have a more difficult time soothing themselves, falling asleep, and staying asleep. That some children take longer to sleep through the night independently is in part a reflection of their inborn tendencies as well as how they are raised.[64]

WHAT ABOUT SLEEP TRAINING?

When Cee was a baby, I didn't know most of what I've written in this chapter. I didn't know that learning to self-soothe could be a gradual process and that it could start early in life. We followed the advice of many mainstream parenting books: do whatever works to soothe your baby for about the first three to four months; then, if your baby is having trouble sleeping, consider sleep training.[65] Sleep training means asking your baby to fall asleep more independently and almost always involves some crying as she gets used to the change.

What worked to soothe newborn Cee was bouncing on an exercise ball. Within a few weeks, we were bouncing Cee to sleep for nearly every single sleep, including the middle of the night, and she was requiring longer and longer stints of bouncing to get settled into a deep sleep. At 3 months of age, she had little ability to self-soothe—it's something she never had a chance to try—and my aching back was telling me that we had picked an unsustainable soothing method. I started trying to rock her to sleep. She screamed at me. She wanted to be bounced, and she didn't know how to settle without it. I knew that she had to learn to sleep in another way.

At this point, I finally had to admit that my presence wasn't helping Cee in her struggle to fall asleep. She needed a little space to learn to sleep on her own. And so we sleep-trained. For us, that meant putting Cee down in her crib after our usual bedtime routine, saying goodnight, and leaving the room. We returned at intervals of a few minutes to give her a little reassurance. She cried, but no more than she had while falling asleep in my arms in the rocking chair. And by the third night, she was fussing for just a few minutes before falling asleep. She had also adopted a lovey, a pink fabric doll that had snuggled with us as we breastfed for the past several weeks. Over the course of a few days, Cee went from being a baby who struggled to fall asleep, despite all our active soothing, to one who went to sleep easily on her own and woke only once in the night to nurse. (She continued to have one nighttime feeding until she dropped it on her own, around 8 months.) She went from having fragmented nighttime sleep of about 8 hours to sleeping 12 hours a night. We would often wake to the sound of her babbling contentedly in her crib after a good night's sleep, whereas before, our mornings had begun abruptly with the sound of her crying. I was finally getting some sleep, too, and I no longer had that horrible feeling of resentment that sometimes crept into my heart when I was bouncing Cee at 2 o'clock in the morning, my back and neck aching. I was well rested and able to be a responsive, sensitive mom, day and night.

This was our story. I'm not proud of the fact that I let my baby cry before sleep when she was so young. In hindsight, I wish that her path toward self-soothing had been more gradual and that I'd been able to support her more along the way. If we have another baby, I hope that the strategies I've described in this chapter will help us find a better start with sleep. But all of that said, I know that sleep-training Cee helped her, and us, to get the sleep that we needed, and it made our nights together more peaceful and sweet. Like every baby, Cee went through later periods of night waking when she was teething and learning to walk. If she woke and cried in the night, we comforted her. But by and large, her ability to self-soothe gave our family the gift of good sleep, and for that I am thankful.

Some people strongly oppose the idea of letting babies cry as they learn to fall asleep on their own. Since sleep-training Cee and writing about

⁂ SLEEP TRAINING AND OTHER BIG CHANGES

* Find a strategy that works for you. You might use a book to guide you or talk with an infant sleep consultant. Make sure you feel comfortable and confident in your new strategy and that it fits your parenting style. The goal is to help your child move toward more independent sleep, usually falling asleep on her own without your active soothing. How you get there is up to you. It can be relatively quick or very gradual, perhaps starting with sitting right next to your baby's crib as she falls asleep and then gradually moving your chair farther away. Consider your child's temperament and your tolerance for her protests. There's not a magic formula; it's different for every family.
* Prepare your child for the change. Explain that she'll be learning to go to sleep in a different way. Tell her that you know that she can do it, and you'll support her as she learns. Even a baby, who may not understand the meaning of all your words, might understand from your tone that a change is coming.
* Be consistent. Once you decide on a strategy, be ready to follow through. Know that it will be hard, but you've thought carefully about this, and you know that your family needs a change. Being confident and consistent will help your baby feel more secure in this change and allow her to adapt more quickly.
* Find support. Talk to a friend who has been through this process, or join a supportive online forum. Find support from parents who understand your philosophy, so they can help you troubleshoot the process without judgment.

it on my blog, I have been told that it was cruel abandonment and that it may have damaged her brain or our relationship. That response made me feel horrible and defensive, but it also compelled me to find all of the research I could on sleep training. I wanted to know, truthfully, how effective it is and whether it might be harmful to babies. Here's a quick summary of what I learned.

1. *Sleep training works.* There is an array of "methods," some gradual and some more abrupt. They all involve letting your baby work on self-soothing so that she can eventually fall asleep without your help, and they all are effective. A 2006 review summarized 11 randomized controlled trials of sleep training, including a total of 1,135 children between the ages of 6 weeks and 5 years, and several more have been conducted since.[66] Together, they find that sleep training leads to reduced bedtime struggles, fewer night wakings, longer sleep for both baby and parents, better maternal mental health, and even improved baby temperament and mood.
2. *But, sleep training doesn't work for every baby.* Dr. Jodi Mindell, associate director of the Sleep Disorders Center at Children's Hospital of Philadelphia, has reviewed just about every study on sleep training,[67] and she herself has published more than 50 peer-reviewed studies on infant sleep. When I asked her if sleep training works for every baby, she replied: "Of course there will be babies who do not respond. Just like with anything—most do well, some receive no benefit, and a few will do poorly."[68] You are more qualified than anyone else to venture a guess at whether sleep training will work for your baby and to know when to throw in the towel if it doesn't seem to be working. And even a baby who knows how to self-soothe will probably go through periods of night waking again. For most families, sleep training improves sleep, but it isn't a magic bullet.
3. *There is no evidence that sleep training will hurt your child.* Those who think sleep training is harmful will cite scientific studies to back their assertions.[69] But if you read those studies, you'll find that they aren't about sleep training; they're about babies who were subjected to chronic neglect or abuse or were raised in orphanages, lacking strong attachment figures. Or they're about nonhuman primates or rodents separated from their mothers for extended periods of time. These are examples of chronic, toxic stress.[70] They're deeply saddening, but they don't tell us much about sleep training in the context of a loving family. Anyone who claims that sleep training will cause long-term harm is not representing the science accurately. And they're ignoring the fact that chronic sleep deprivation could be harming you and your baby. Sleep deprivation increases mom's risk of postpartum depression,[71] may add to relationship stress,[72] and may make it dangerous to drive

a car,[73] among other things. And you will make up for a few nights of tears with all of the positive parenting interactions that come with a well-rested family.

4. *Sleep training is an imperfect solution.* Really. As much as I know it helped our family, it's still hard. It's a big change, and it is probably stressful to babies.[74] It is stressful to parents as well, particularly if half of your friends are telling you that it is a Big Mistake that only Bad Parents make. I'd like to think that setting your baby up with good sleep habits

✲ RECOMMENDED SLEEP RESOURCES

I have yet to find the perfect sleep book, but that's because it would be impossible for even the smartest baby sleep guru to write the book just for you and your baby. I like all of the books listed here for different reasons, but I recommend that you spend some time finding one that feels like a good fit for you. You might consider checking several out from the local library and skimming through them to see which ones resonate with you.

The Happy Sleeper: The Science-Backed Guide to Helping Your Baby Get a Good Night's Sleep—Newborn to School Age by Heather Turgeon and Julie Wright (2014)
Bedtiming: The Parent's Guide to Getting Your Child to Sleep at Just the Right Age by Isabela Granic and Marc Lewis (2010)
Healthy Sleep Habits, Happy Child by Marc Weissbluth (1999)
Sleeping through the Night: How Infants, Toddlers, and Their Parents Can Get a Good Night's Sleep by Jodi Mindell (2005)
The Sleep Lady's Good Night, Sleep Tight: Gentle Proven Solutions to Help Your Child Sleep Well and Wake Up Happy by Kim West (2009)
Solve Your Child's Sleep Problems by Richard Ferber (2006)

If you're struggling with sleep, it may also be helpful to hire an infant sleep consultant who can help you come up with a plan that is personalized for your baby and your parenting style. A directory of sleep consultants, with their credentials, can be found at www.iacsc.com.

from the start can help the transition to independent sleep be more gradual and gentle. When that doesn't work, sleep training is one solution, but it isn't the only solution. Bed sharing may work in some families. Having your partner take over more of the nighttime parenting might help in others. Each family is different in what works for them.

5. *You don't have to abandon your baby in a dark room*. And you don't have to follow any rules or sign onto a specific method. You can follow your heart for setting gentle boundaries for sleep. It can be a gradual and supportive process—whatever works best for you and your baby.

As you work through sleep struggles with your baby, try to remain open-minded, to observe and respect your baby's needs and abilities, and to work within your own zone of comfort and your parenting philosophy. Tom Anders gave me this wise warning: "I think that parental culture, belief systems, and values are really important ingredients to take into consideration. Good sleep is promoted by a calming, secure, comforting environment, and if parents are trying to impose some kind of a behavioral technique that isn't consonant with their own belief system, I don't think that's very helpful."[75]

Whatever you do, do it mindfully, lovingly, and respectfully. And then, please, don't feel guilty about your choice. If you feel judged by others, remember that they don't live in your house at night, and they don't care for your child. You do, and you are capable of doing the right thing for your child. You and your family need sleep. And if you find that whatever you're doing isn't working for you, don't be afraid to change course. There is no overwhelming evidence that any one choice in nighttime parenting is superior to another. The one thing we know is that if your choice is working for you, it is superior for your family.

8 ⁂ VACCINES AND YOUR CHILD

Making a Science-Based Decision

Soon after they were married, my paternal grandparents had three boys in quick succession. The first was my father, Richard, born in October of 1948. Next came Frankie, 18 months later. And Larry was born less than a year after Frankie. Three boys in the span of three years, they were close in age and close friends.

In the black-and-white photographs from their childhood, it is Frankie's face that I study the closest. Frankie died when he was 6 years old. My father never talked about him, so I know him only through photographs. He is smiling in nearly every one.

As a child, I never thought very much about Frankie. I figured that the death of a child wasn't unusual during the "olden days." My grandparents had seven children in all, and in my childish mind, it seemed like that might compensate somehow for the loss of Frankie. But then I became a mother myself, and when I think about losing Cee, I feel like my heart might explode. I told my grandmother this as we sat down to talk about Frankie, and she nodded. "Each one is special," she said.[1]

My grandmother, Margaret Green, is just a few months shy of 90 years old as I write this. Her mind is sharp and her hugs are firm. She is the only person on earth who can tell me anything about Frankie. My grandfather passed away almost 25 years ago, and my own father died a few years later. Margaret's small apartment is packed with shoeboxes full of photographs and old letters, but most of the

Three brothers (from left to right): Richard, Frankie, and Larry Green, circa 1953 or 1954, in Princeton, New Jersey. Frankie died in 1956, at age 6, of encephalitis caused by measles. Photo by Margaret Green, used with permission.

memories of Frankie are carried close to her heart. So we sit and talk about him, a little boy who never grew up, the uncle I never met.

My grandmother remembers how Frankie adored his older brother, my father, and how one day they played hooky from school together, entertaining themselves in the town of Princeton, New Jersey. They must have been in first grade and kindergarten then, and according to my grandmother, they got in "mild trouble" for this mischief, but really she was grateful for her boys' spirit and close friendship. She remembers how desperately Frankie wanted a pair of toy six-shooter pistols the Christmas before he died, and how when she suggested that the right-shaped stick might work instead, he sweetly shook his head. "No, Mom, I don't think so," he patiently explained. My grandparents gave Frankie his six-shooters that Christmas, and Margaret is thankful they did. And she remembers Frankie riding a merry-go-round for the first time, beaming at her with each turn. "I can still see his face, and that happy, happy smile."

Margaret also remembers when all three boys came down with measles in May of 1956. How lucky, the neighbors said, for the boys to get measles at the same time. In those days, measles was a rite of passage, a part of childhood. Nearly every child suffered through it at some point, but once they had, they would be immune for life. Parents often intentionally exposed their children at "measles parties" so that the whole playgroup would get the disease over with at once. It wasn't hard to infect a group of children quickly, intentionally or not. Measles is one of the most contagious pathogens on earth. Those infected are contagious for several days before and after the characteristic rash appears. With every cough or sneeze, the virus flies around in airborne droplets, where it can survive for two hours.[2]

Having measles meant being stuck at home for a couple of weeks, at first with symptoms that seemed like a common cold: cough, runny nose, fever, and often an eye infection. Several days later, measles clearly announced itself as an itchy rash that spread all over the body.[3] There was no doubt that this was uncomfortable, but, my grandmother told me, "It was also a special time." The boys got ice cream and presents, and she read them lots of books. "And the three of them were together," she said.

After a week or so, the boys appeared to be getting better, their rashes subsiding. My grandfather had just been offered a job at Johns Hopkins University, and he drove to Baltimore to look for a place for the family to live. Margaret continued nursing the boys back to health on her own. She vividly remembers the night when Frankie's case of measles took a turn for the worse. "I settled all three boys down to bed. These were sick boys, but they were recuperating quickly, and they were all on their way to being completely well. I got Frankie settled and got him his glass of water and then went to bed myself. But as I was getting ready for bed, I heard this kind of funny noise and went in just to check one last time. He was draped halfway out of the bed, which was strange, and I rushed over to pick him up and get him settled and realized that he was unconscious."

Frankie never woke up. He suffered from one of the cruelest complications of measles: encephalitis, or inflammation of the brain. Sometimes it is a primary infection, occurring at the same time as the rash phase of the illness. The measles virus invades the brain and replicates there, directly damaging neurons and causing brain swelling and inflammation. Frankie probably died of a second type of measles encephalitis, which appears when the patient seems to be on the mend. In this case, the encephalitis is characterized by damage to the myelin sheath that coats nerves and allows conduction of nervous signals. Both types of encephalitis are very serious, even with the best modern medical care. Those who survive often suffer permanent brain damage, hearing loss, and seizure disorders.[4]

As Margaret remembers it, Frankie was hospitalized, in a coma, for about a week before he died. Finally, after an urgent call from the hospital, she left the other children with a neighbor and hurried to see her boy. The pediatrician met her at the door to say that Frankie had already died. Then the doctor took Margaret to her own home and put her to bed in her guest room to let the news sink in before she went back to Richard and Larry and tried to explain where their brother had gone.

It's an unfathomable loss, and although it came as a shock to my grandparents and Frankie's brothers, his death, sadly, wasn't unusual. Frankie's was one of more than 600,000 reported cases of measles in the United States in 1956.[5] Actual numbers were probably much higher, thought to reach three to four million cases per year, since many went unreported. In the decades before the introduction of the measles vaccine, about 48,000 people were hospitalized with measles each year; 7,000 had seizures, 2,000 suffered permanent deafness or brain damage, and 500 died.[6] Every year. In 1956, Frankie was one of 530 lives lost to measles in the United States, most of them children.[7]

There are other, more common, complications of measles. One in 20 people infected get pneumonia. One in 10 get an ear infection. As Dr. C. Everett Koop, pediatric surgeon and former U.S. Surgeon General, recalled, "Children would have measles . . . and they would develop an infection in their ear . . . Out of their ear would drip, for years, a foul-smelling gray-green pus, and it lasted until some of them were adults."[8]

Measles was a routine part of childhood, but that didn't make it any less scary. And had Frankie been born just a few years later, his death might have been prevented by the measles vaccine.

When my grandmother hears that my generation of parents is worried about vaccines, she thinks about Frankie. She also lived through the polio outbreaks that crippled children and the rubella outbreaks that caused thousands of miscarriages, stillbirths, and birth defects. She has a hard time understanding why parents today hesitate to vaccinate their kids. And I think that we can't discuss the current vaccine controversies without first acknowledging the history of parenting under the dark cloud of deadly childhood diseases, just a couple of generations ago.

DEVELOPMENT OF THE MEASLES VACCINE

Even as Frankie fought and lost his bout with measles, scientists were hard at work on a vaccine. Harvard professor John Enders headed an infectious disease laboratory at Children's Medical Center in Boston. Dr. Enders was a brilliant virologist, and he was also a generous humanitarian.[9] He never tried to patent his work, and he believed that vaccine development benefited from shared knowledge and collaboration with fellow scientists. In 1954, he and two colleagues were awarded the Nobel Prize in Physiology or Medicine for their work on growing poliovirus in cell culture, techniques that led to the development of the polio vaccine.[10] While this was a great accomplishment, Enders was actually more interested in measles. He considered it the more pressing problem, since, worldwide, 8 million children were dying of measles each year.[11]

So Enders and his colleagues set out to collect samples of the measles virus from boys at a boarding school in Boston, where close living quarters meant frequent outbreaks. The scientists finally isolated a strain of measles swabbed from the throat of an 11-year-old boy named David Edmonston, and this same strain led to the measles vaccine.[12]

First, the scientists worked on getting the isolated measles virus to grow in human cell cultures, but to make a safe vaccine, they needed to attenuate, or weaken, it. Since humans are the only natural host for the measles virus, they started growing their strain of measles in chick embryo cell cultures. In this environment, the virus had to change to survive, making it progressively weaker in humans. After three years of work, the attenuated Edmonston strain of measles was ready to test for its potential as a vaccine. It was first tested in monkeys, and then Enders and his research fellows tested it on themselves. They all showed a strong antibody response and had no adverse effects. Subsequent tests on children showed that they were completely protected in the next outbreak, although many developed a high fever and rash in response to the new vaccine.[13]

True to his nature, Enders shared the vaccine material with scientists around the country and encouraged them to work on refining the vaccine and getting it into production. Maurice Hilleman, who headed virological research at Merck Pharmaceuticals, was ultimately responsible for developing the measles vaccine in use today. When he started working with Enders's vaccine material, he considered it "toxic as hell," given the high

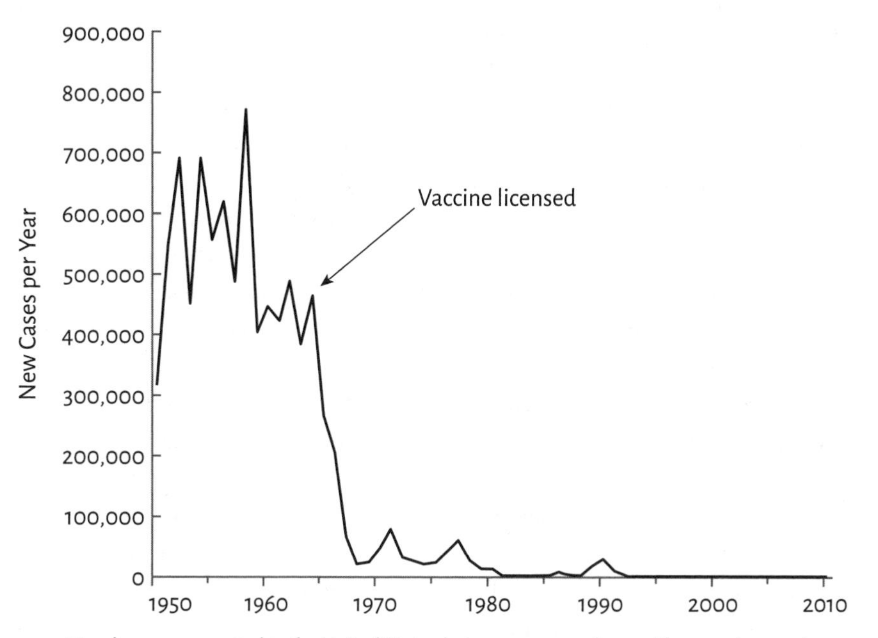

Measles cases reported in the United States between 1950 and 2010. The measles vaccine was first licensed in 1963. *Source:* National Center for Health Statistics, *Health, United States, 2012: With Special Feature on Emergency Care* (NCHS, 2012); Centers for Disease Control and Prevention, *MMWR Summary of Notifiable Diseases, United States, 1993* (CDC, 1994).

incidence of side effects.[14] Hilleman felt pressured to get the vaccine out quickly, but he insisted on taking the time to make several modifications to reduce side effects and ensure the vaccine was safe. His version of the vaccine was first licensed for use in the United States in 1963, and a further attenuated version followed in 1968. In 1971, the measles vaccine was incorporated into the MMR combination shot, which also includes vaccines for mumps and rubella.[15]

The measles vaccine was remarkably successful. Two doses of MMR induce immunity to measles in 99% of people, and after release of the vaccine, the incidence of measles dropped dramatically.[16] By 2000, measles was nearly eliminated from the United States. In the decade that followed, there were about 60 cases per year, almost always traced to international travel.[17] The virus would enter the United States with an infected individual, who was usually unvaccinated and unaware that he was carrying the disease, since the early symptoms can seem like a common cold. But with most of the population vaccinated, measles couldn't spread very far, and outbreaks were rare and easily contained.

In the past couple of years, however, we've witnessed a disturbing trend of larger and more frequent measles outbreaks around the United States. With more parents opting out of immunizing their kids, clusters of unvaccinated people have allowed the disease to spread. In 2011, there were 222 cases of measles in the United States; in 2013, there were 189. As I write this, 2014 is already the worst year we've seen in two decades, with nearly 600 cases reported by mid-August.[18] There seems to be a report of a new outbreak each week.

The irony is that as some American parents are choosing not to vaccinate their kids, children in other parts of the world continue to suffer due to lack of vaccines. Worldwide, measles still killed 122,000 people in 2012, and that number was down from 562,000 in 2000 due to a concerted effort to get the MMR vaccine to more children. About 98% of measles deaths occur in Africa, the Middle East, and South-East Asia, where the percentage of children receiving the first dose of measles vaccine lies between 50% and 75%.[19]

HOW VACCINES HAVE CHANGED CHILDHOOD

Frankie died of measles, but his brothers fully recovered. During the course of their infections, their immune systems responded by developing immune cells that produced antibodies to the measles virus. These cells would stay around for the rest of their lives, ready to respond to another exposure to measles. Of course, this immunity came at a price, given the high incidence of complications of the disease, which my dad's family knew all too well.

The measles vaccine also induces immunity to the measles virus, but it does so in a much gentler way, without the risks that come with a full-blown infection. Vaccines contain components of the infectious bacteria or virus, weakened in some way. Our immune systems still recognize the pathogen and produce antibodies to fight it, but it doesn't make us sick. A vaccine effectively educates our immune systems about a particular pathogen so that if we are exposed to the disease in the future, we're ready to respond. With a vaccine, we not only get immunity without getting sick, but we also get to control the amount of exposure to the pathogen (using the smallest dose possible) and the timing of exposure, aiming to induce immunity as soon as a child's immune system can safely and effectively do so. This is why we give babies so many vaccines—because infections are usually most dangerous in the first years of life. If we waited until later in childhood to vaccinate our children, we would be leaving them unprotected from diseases during a very vulnerable time.

Measles is but one of at least 14 diseases that your child may be vaccinated against in the first two years of life.[20] As each vaccine was approved and recommended for children, there was a subsequent drop in the numbers of both cases and deaths, most reaching a 99% to 100% decrease compared with pre-vaccine levels.[21] For these diseases, the historical numbers of cases and deaths—each one representing someone's beloved child—are so large that they are incomprehensible to us today. Each year in the United States, the modern childhood vaccination schedule is thought to prevent nearly 20 million cases of disease and 42,000 deaths and to save $69 billion in direct medical costs and indirect costs to society (due to missed work, lost productivity, and disability).[22]

Sometimes, people who question the value of vaccines point out that prior to vaccines, the death toll from infectious diseases had already been dramatically reduced due to improved nutrition, sanitation, clean water, antibiotics, and advances in medical care. This is true. However, all of these advances still left an awful lot of people suffering through these diseases and their complications, and although modern medicine and public health measures may have improved the chances of survival, previously healthy kids like Frankie were still dying.

We also have examples of vaccines that have been introduced in just the past few decades, like varicella (chickenpox) in 1995 and rotavirus in 2006. There haven't been any overwhelming advances in sanitation or medicine in the United States during that time, but these vaccines have clearly reduced illness and death in children. Before the varicella vaccine, the disease caused an average of 11,000 hospital admissions and 103 deaths per year in the United States. Since the vaccine was introduced, these numbers have decreased by 90%.[23]

Rotavirus can cause severe diarrhea and vomiting in babies and young children. It doesn't discriminate; regardless of income, hygiene, nutrition, or water source, it infects nearly every child in the world within the first years of life (although severity of symptoms varies). In developing

✲ HISTORICAL CHANGES IN VACCINE-PREVENTABLE DISEASES IN THE UNITED STATES, PRE- AND POST-VACCINE

	Pre-vaccine Annual Average		*Post-vaccine Reports or Estimates, 2006*		*Pre-Post % Reduction*	
DISEASE	*CASES*	*DEATHS*	*CASES*	*DEATHS*	*CASES*	*DEATHS*
			Vaccines recommended before 1980			
Diphtheria	21,053	1,822	0	0	100%	100%
Measles	530,217	440	55	0	100%	100%
Mumps*	162,344	39	6,584	1	96%	97%
Pertussis	200,752	4,043	15,632	9	92%	>99%
Poliomyelitis, paralytic	16,316	1,879	0	0	100%	100%
Rubella	47,745	17	11	0	100%	100%
Smallpox	29,005	337	0	0	100%	100%
Tetanus	580	472	41	4	93%	99%

Vaccines recommended between 1980 and 2014						
Hepatitis A	117,333	137	15,298	18	87%	87%
Hepatitis B, acute	66,232	237	13,169	47	80%	80%
Invasive Hib†	20,000	1,000	<50	<5	>99%	>99%
Invasive pneumococcal disease‡	63,067	6,500	41,550	4,850	34%	25%
Varicella	4,085,120	105	612,768	18	85%	83%

Sources: S. W. Roush, T. V. Murphy, and Vaccine-Preventable Disease Table Working Group, "Historical Comparisons of Morbidity and Mortality for Vaccine-Preventable Diseases in the United States," *JAMA* 298 (2007): 2155-63; W. Atkinson, S. Wolfe, and J. Hamborsky, eds., *Epidemiology and Prevention of Vaccine-Preventable Diseases*, 12th ed. (Washington, DC: Public Health Foundation, 2012).

*2006 was an unusually bad year for mumps due to outbreaks in college dormitories. Most years, there are only about 20 mumps cases.

†Haemophilus influenzae B; post-vaccine estimates are for 2005.

‡Post-vaccine estimates are for 2005.

countries, where children often don't have access to good medical care, rotavirus is thought to kill half a million children each year.[24] In the United States, modern medicine saves most babies from dying of rotavirus, but before the vaccine, rotavirus still caused more than 600,000 doctor and emergency room visits, 67,000 hospital admissions, and 30 deaths per year—a heavy burden. By 2010, there was a 96% decrease in the number of hospitalizations for rotavirus infections in U.S. children under 2 years of age.[25] Improvements in hygiene or nutrition didn't prevent these illnesses; the vaccine did.

The success of vaccines means that many previously common diseases are now rare, but even with less disease circulating in our communities, unvaccinated children remain vulnerable. For example, an unvaccinated child is 60 times more likely to contract measles and up to 20 times more likely to catch pertussis (whooping cough) than a vaccinated child.[26] Communities with more unvaccinated kids are more likely to see outbreaks of vaccine-preventable diseases.[27] In the unlikely case that a vaccinated child still gets measles or pertussis (because no vaccine works 100% of the time for every person), studies show the illness will be less severe, of shorter duration, and less likely to result in hospitalization.[28]

Would it be more natural to let children fight off these diseases on their own, without the protection of vaccines? Maybe, but that's an awfully cruel side of nature to entertain. It also isn't natural to fly around in airplanes, carrying diseases around the globe, or to congregate in tight spaces like commuter trains, shopping malls, or indoor playgrounds, and this unnatural behavior puts us at constant risk for outbreaks. Without vaccines, more healthy children like Frankie would be dying, and we would spend much more of our time and energy nursing sick kids, quarantining them to protect the more vulnerable, and worrying about whether they'd survive the latest bout of illness. We'd use more antibiotics, with increasing rates of resistance. We'd be nursing more sick kids in NICUs and PICUs, trying to keep them hydrated, nourished, and breathing, and billing insurance companies or taxpayers for each day in the hospital. This model of handling disease is far more costly in lives, suffering, and dollars (and far more profitable for pharmaceutical companies) than the model of preventive medicine.

With vaccines, we are intentionally exposed to germs in a controlled, well-tested way, strengthening the immune system without the risk of the "natural" illness. We know that not vaccinating would take us back to the days of countrywide outbreaks and encephalitis sneaking in to steal away a little boy just after bedtime. Maybe that's more natural, but it sure isn't better.

HOW VACCINES PROTECT COMMUNITIES

High rates of vaccination protect our communities from the spread of disease. If most of the people in a population are vaccinated, then the odds are low that a single infected individual will infect others and cause a large outbreak. This is called herd immunity. It's important because it is how we protect people who can't be vaccinated, including infants too young to receive vaccines, people who are allergic to vaccine ingredients, and those who are immunocompromised.

For each vaccine-preventable disease, we can estimate a threshold for herd immunity, or the percentage of the population that must be vaccinated to protect it from an outbreak. This threshold depends on how quickly the disease can spread from one person to another, how well the vaccine works (a small percentage of people won't respond to it), and how the unvaccinated are distributed in the population. If there are pockets with a lot of unvaccinated individuals, the chances of an infection spreading are greater.[29] For most diseases, we need about 75% to 85% of the population to be vaccinated to minimize outbreaks and protect susceptible individuals. Measles and pertussis have higher thresholds for herd immunity, greater than 90%, because they spread so easily.[30]

In 2008, a 7-year-old unvaccinated boy returned from a trip to Switzerland to his home in San Diego, his parents not realizing he was infected with measles.[31] If this little boy had encountered only people who were immune to measles, as 99% of vaccinated folks are, then he might have been the only person affected. Instead, his unvaccinated brother and sister were immediately infected, and then, between the three of them, they went to school, dance class, their pediatrician's office, and finally the emergency room. Several unvaccinated classmates were infected, as were children in the waiting room of the doctor's office—three infants too young to be vaccinated and one 2-year-old whose parents had decided to delay the MMR shot. Other children infected in this outbreak went to swimming pools, grocery stores, indoor playgrounds, and the circus. In these everyday activities, hundreds of people were exposed. Twelve children were ultimately infected with measles, all of them unvaccinated, and one infant was hospitalized.[32]

The San Diego boy's immediate community, including many families who had chosen not to vaccinate their children, had low herd immunity, and that's what allowed this outbreak to spread as much as it did. The larger San Diego community had good vaccine coverage, with 97% of children receiving the MMR. Between high vaccination rates and the 21-day quarantine of infected kids, this outbreak couldn't spread further.[33] The story was different in 1989-91 when the United States experienced large outbreaks of measles, resulting in 55,000 cases and 136 deaths, mostly in unvaccinated preschoolers. At that time, vaccination rates in preschool-aged children were between 60% and 70%, far below the threshold for herd immunity.[34]

We also get to see herd immunity at work when we introduce new vaccines. For example, the varicella vaccine is given between 12 and 15 months of age. For the first few months of life, maternal antibodies that passed across the placenta during pregnancy provide some protection to the baby, but when that wears off, babies are susceptible to varicella until they can be vaccinated. After introduction of the varicella vaccine in 1996, the incidence of varicella in *unvaccinated* infants less than 12 months old decreased by 90%.[35] These babies were benefiting from indirect protection by the vaccine, because there was less varicella virus circulating around them. Likewise, when we began immunizing infants for rotavirus, rates of rotavirus infection dropped in *unvaccinated* children and even in adults.[36] The vaccine was directly protecting babies, and the people around them were benefiting indirectly, through herd immunity.

These are just a couple of examples. Herd immunity is happening all the time, with nearly every vaccine, in our immediate families and communities.[37] Of course, herd immunity also protects children whose parents choose not to vaccinate them because of personal or philosophical beliefs. This is a good thing; nobody would wish a preventable disease on innocent children. But it's also a dangerous gamble for these families, and a community can sustain only so many of these "free-riders" (that's the academic term) before it is susceptible to a serious outbreak, threatening not just those who are unvaccinated by choice (or rather, their parents' choice) but everyone who doesn't have immunity to the disease, including young babies and those unable to be vaccinated for medical reasons. In a paper on herd immunity, researchers aptly commented: "A single unvaccinated child in a community of vaccinated children holds a strategically opportunistic high ground, protected from risk of disease by herd immunity while avoiding risk of exceedingly rare adverse events associated with vaccination. Yet, when too many parents want their child to be that child, the entire community is affected."[38]

Parents who choose not to vaccinate their children might point out how naturally healthy they are, but they forget that the health their children enjoy rests on the protection provided by herd immunity. Vaccinating your child means contributing to the collective health of your community and protecting those more vulnerable.

UNDERSTANDING VACCINE RISKS

Although the benefits of vaccines are tremendous, they do carry some risk. But then again, nothing we do is 100% safe. Every time we feed our children, we take a potentially fatal risk (choking, food-borne illnesses, allergies, etc.), but we know that the benefits of eating far outweigh the risks. For each vaccine we give to our children, we want to understand the risks, ensure that they're smaller than the benefits, and minimize them as much as possible.

The risks of side effects or reactions from vaccines are well understood. They're listed on the CDC website and on the information sheets given to parents with each vaccine their child receives. And for the vast majority of people, the benefits of vaccine protection are much greater than the risks.

Take the MMR shot, for example. We've been using MMR for more than 40 years, and we have data from more than 60 studies, including 15 million children, to help us understand its risks. There are common but mild side effects, including fever, occurring in 1 in 7 children, and a short-term rash in 1 in 20. There are rarer side effects, such as febrile seizure in 1 in 3,000 to 4,000 MMR recipients and a temporary blood clotting disorder in 1 in 40,000.[39] Febrile seizures can occur any time a child has a fever, and they aren't thought to be dangerous—although they can be very scary to parents.[40] However, these side effects are miniscule compared with the risks of the diseases themselves: for measles, a 1 in 20 chance of pneumonia and 1 in 1,000 chance of encephalitis; for mumps, a 1 in 7 chance of meningitis and, in men, a 1 in 2 chance of testicular inflammation, often leading to atrophy; and for rubella infection during the first 20 weeks of pregnancy, almost certain harm to the fetus, causing miscarriage or major birth defects.[41]

Most other vaccines have only mild side effects. Very rarely, a child will have a severe allergic reaction to a vaccine component; the chances of this are about one in a million.[42] But of course, every time your baby tries a new food or medicine, there is a risk of an allergic reaction.

Still, the risks and benefits of vaccines can seem so intangible that it can be daunting to try to tally them up on our own. A great resource is the Institute of Medicine (IOM), a nonprofit organization that works outside the government to bring together unbiased, interdisciplinary, volunteer experts to assess health concerns. Its reports—each hundreds of pages—are available to the public online, and they've assessed a number of questions regarding vaccines, including concerns about links with autism, SIDS, and immune dysfunction, and ingredients in vaccines.[43] In 2013, an IOM committee, specifically selected to exclude anyone who had financial ties to the pharmaceutical industry or had previously served on a federal vaccine committee, assessed the safety of the entire childhood immunization schedule; it found no evidence that the current schedule isn't safe.[44]

What the IOM does is systematically review all the research on vaccines to reach a scientific consensus. This is vital, because among the thousands of scientific publications on vaccine safety, there *are* a few concerning reports, and these will often be cherry-picked by anti-vaccine activists to build a story about the dangers of vaccines. Sometimes they're preliminary studies, sometimes their methodology is shoddy, and sometimes they're simply misinterpreted. In their reviews, the IOM committees include these papers but put them in context with the rest of the research to arrive at a scientific consensus.

In the appendixes of this book, I address several specific vaccine safety concerns, including the worry that we give too many too soon (appendix C), links with autism (appendix D) and SIDS (appendix E), aluminum in vaccines (appendix F), and why the hepatitis B vaccine is recommended at birth in the United States (appendix B).

VACCINE TESTING AND SAFETY MONITORING

What about when a new vaccine is introduced? How do we know that its benefits will outweigh the risks? Let's walk through this process by looking at the history of the rotavirus vaccine, the most recent addition to the childhood schedule.

Globally, rotavirus kills nearly 500,000 children each year, and developing a vaccine has been a research priority since the 1970s. That research started in labs at universities and research institutes, where scientists worked out the intricate details of how the virus worked and developed potential vaccine strains. By the 1990s, the leading vaccine candidate, called RotaShield, was being tested by the pharmaceutical company Wyeth.[45]

Before the Food and Drug Administration allows a new vaccine to be tested in humans, the pharmaceutical company first has to show that it is safe and effective in animals. Then it is tested in three phases of human clinical trials, the third phase including thousands of people tracked for several years. Any new childhood vaccine must be tested in children along with the rest of the immunization schedule to ensure that it works and is safe not only on its own but also in the context of other vaccines. If the trials go well, the company submits its data to the FDA.[46]

If the FDA approves a vaccine, the CDC works with an external group of advisors, the Advisory Committee on Immunization Practices (ACIP), to determine whether it makes sense to give the vaccine in the United States, given the prevalence and seriousness of the disease, and at what age it should be given. ACIP reviews vaccine recommendations every year, not just looking at new vaccines but also monitoring any new studies on established vaccines.[47]

In early 1999, ACIP recommended that RotaShield be added to the immunization schedule for babies.[48] At this point, clinical trials including more than 10,000 babies showed that the vaccine worked very well and had an excellent safety profile. However, Wyeth, the FDA, and ACIP noted that a few infants had developed intussusception soon after getting the vaccine. This is a serious but treatable problem in which one part of the intestine slides into another part, like sections of a telescope. In the clinical trials, the intussusception cases appeared to be coincidental, but everyone would be keeping an eye on this when the vaccine was introduced.

After a vaccine is introduced, one way that its safety is monitored is through the Vaccine Adverse Event Reporting System (VAERS).[49] VAERS allows anyone—usually doctors or parents—to submit reports of any outcome that is suspected of being related to a vaccination. The CDC and FDA watch VAERS for patterns that seem unusual, recognizing that many reports aren't necessarily linked by causation but could be coincidences. Within a few months of the release of RotaShield, VAERS had received 12 reports of intussusception. The CDC immediately began an investigation, and it suspended use of RotaShield in July of 1999. By October, Wyeth had taken it off the market. After the dust settled, it was estimated that RotaShield caused intussusception in about 1 in 10,000 vaccinated babies.[50]

The example of RotaShield illustrates a couple of things. First, the clinical trials and the multiple agencies that review safety and efficacy data serve a vital role, but they aren't foolproof. A 1 in 10,000 risk won't show up in a clinical trial of 10,000 children, but at least we know that any risk not identified at that stage is very rare. Second, after RotaShield was approved, the problem of intussusception was quickly detected through VAERS, and the CDC acted quickly. The system worked.

This didn't solve the problem of rotavirus, however. Without a vaccine, rotavirus would continue hospitalizing tens of thousands of babies in the United States. RotaShield's demise also had a significant global impact. Once U.S. officials said that it wasn't safe enough for American children, it didn't have much of a chance internationally, where it might have provided much greater benefit even with the risk of intussusception. "I'd love to use it, we have 100,000 deaths each year of this disease. But when I have the first case of intussusception I will be tarnished in the press for having accepted a vaccine that was rejected in the U.S.," one senior Indian health official said.[51]

Two more rotavirus vaccines were developed after RotaShield. This time the clinical trials were much larger, including more than 60,000 infants, to be more certain about intussusception risks. The newer vaccines showed no increased risk of intussusception and were recommended by the CDC and WHO for use around the world. Still, everyone kept watching for intussusception, and again, vaccine safety monitoring systems were used to carefully track the outcomes from these new vaccines. The CDC runs a program called the Vaccine Safety Datalink (VSD), a network of managed health care systems that tracks data on nine million people.[52] Researchers used the VSD system, along with additional safety data from the FDA, and concluded in 2014 that both new vaccines appear to cause intussusception at a rate of 1 to 5 in 100,000 infants, a lower rate than RotaShield.[53]

Some context for these findings is important. First, the baseline rate for intussusception in the United States is about 30 per 100,000 babies, so most cases are unrelated to a vaccine. Second, although the studies show a slight increase in risk soon after getting the rotavirus vaccine, the overall rate of intussusception hasn't increased with the introduction of any of the rotavirus vaccines.[54] If the vaccine is causing a small number of intussusception cases, then it's possible that natural rotavirus infection is also a rare cause of intussusception, and some babies are just susceptible to intussusception whether triggered by the vaccine or a natural infection.[55] But regardless, since we now have such a large database on rotavirus and these vaccines, we can say with confidence that the benefits exceed the risks. It is estimated that a baby is 1,100 times more likely to be hospitalized because of a rotavirus infection (without the vaccine) than because of intussusception (with the vaccine).[56]

Everyone—you, me, and the CDC—wants to know whether a vaccine has a serious side effect, however rare. The sheer number of acronyms in this section of the chapter is a testament to the multiple regulatory agencies and monitoring systems that do their best to ensure the safety of every vaccine. A lot of data collection and safeguards are happening beneath the surface, and history has shown that when problems arise, they are quickly addressed. If you're worried about adverse effects of vaccines, understand that the CDC is too. It knows it can't afford a major vaccine safety crisis, and none of us can afford to go back to the pre-vaccine era.

WHY WE STRUGGLE WITH THE CHOICE TO VACCINATE

Janie Oyakawa, an occupational therapist, lives in Prosper, Texas, with her husband and their six children.[57] A traumatic childbirth experience seeded her distrust in modern medicine. She found like-minded "crunchy" moms, especially online, who were also renouncing mainstream practices in favor of parenting more naturally. She stopped vaccinating her kids. Her sixth child didn't see a pediatrician until she was almost a year old. Janie wrote on her blog: "I was very proud of that fact. I wasn't necessarily 'anti-vax,' I was just . . . done."[58] And she was worried. She read several books and Internet sources that left her feeling uneasy about vaccine ingredients, a possible link with autism, and the sheer number of vaccines babies get these days.

Janie was not alone in her concerns. Although the vast majority of parents in the United States continue to fully vaccinate their children, most also report being a little worried about vaccines, and only about half feel very confident in their decision to vaccinate.[59] Why do we parents struggle so much with this decision?

I posed that question to University of Oregon psychologist Paul Slovic, who for more than 50 years has studied the science of how we make decisions.[60] He and others have found that there are essentially two ways that we approach decisions: fast and slow.[61] Dr. Slovic explained to me that we rely on the fast system for most of the choices we make throughout our days. "We make most decisions intuitively, on the basis of gut feeling, influenced by thoughts of the outcome, stories we may have heard, images that flick through our minds in a fraction of a second that carry feeling . . . The fast system was adaptive for millions of years over the evolution of humans. It was the way we dealt with risk before we had science."

I asked Slovic about the choice to vaccinate my child. If I relied on my fast system, with all of its evolutionary success, to tell me whether or not I should let a nurse stick a needle delivering some element of an infectious disease into my baby's soft thigh . . . Slovic interrupted me: "You wouldn't do it."

But, he told me, "Over eons of time, the brain evolved to process information in another way—in an analytic, deliberate way, and to create scientific methods to enhance our knowledge." For decisions about vaccinating our children, we really need to call on this slow system. This is the system that helps us to weigh evidence in a rational way and to understand things that we can't see or feel in the moment.

Without science and critical thinking, we are easily led astray by common cognitive biases.[62] For example, we tend to perceive something that is natural as safer than something that is made by humans, even though nature is full of toxins and brutal tragedies. We don't like to take risks that may carry unknown side effects, even if very rare, but we'll willingly take greater risks when we believe we understand and can control the potential for harm. One of the riskiest choices that we make in our daily lives is to ride in a car, but we feel comfortable with this risk because it is familiar, understandable, and, to some extent, controllable.

And then there are the stories. For example, in your search for information about vaccines, you might happen upon a story of a child who was vaccinated and began to show signs of autism soon after. Even if you rationally know that the story might be untrue and that correlation is not the same as causation, it still plants a seed of doubt. "Stories carry a very powerful emotional impact. The perception of risk resides in us as a feeling as much as it does as a result of some analysis of the evidence, but we need to be aware that this feeling can be very misleading," Dr. Slovic told me.[63] On the Internet, it is far easier to find stories of adverse reactions—true or not—than it is to find stories of uneventful vaccinations, which happen in the thousands every single day, or stories of children dying of vaccine-preventable diseases. This inflates our perception of the risks of vaccines while at the same time deflating our perception of the risk of the disease.

Of course, the stories that flickered through parents' minds a few generations ago were quite different. My grandmother's stories were of Frankie and the family friends who had suffered from polio. A generation later, my parents were part of the counterculture movement of the 1960s and 1970s. They questioned authority, distrusted the government, and sought to raise their children naturally. I was born at home and breastfed. My parents raised grass-fed beef cows, tended a huge vegetable garden, and made farm-fresh, homemade baby food using a hand-cranked grinder. But there was no question that we would be vaccinated. My mom doesn't recall so much as a discussion about it. My hunch is that the consequences of these diseases, including the measles that killed Frankie, were still fresh in my parents' minds. But now, the stories of suffering from vaccine-preventable diseases are trapped in history books and lost with our grandparents. Vaccines are victim of their own success. When they work, nothing happens. There's no story.

My parents also didn't have the Internet. If they had, they might have made a different choice. There is just as much misinformation as accurate information on the Internet, and it can be difficult to distinguish between the two. If you're worried about vaccines, you can find confirmation for that concern online, even if it isn't backed by good science. And being human, we tend to judge information as more reliable if it is consistent with our preexisting beliefs rather than to seek out information that challenges our beliefs.[64] As parents, we have to hold a very high standard for quality of vaccine information.

⁎ HOW TO FIND RELIABLE INFORMATION ABOUT VACCINES

If you search for vaccine information, you will find a mix of reliable and unreliable sources. The Internet is an equal-opportunity platform: anyone can put information there without backing it up with good science. Books aren't necessarily better, because anyone can self-publish or find a publisher that is concerned with book sales, not accuracy. It's really up to you to figure out whether you're looking at a reliable source or have fallen into a misinformation trap. Here are some tips for finding good information about vaccines.

Look for sources that . . .

- Have authors with some credentials—such as an MD or PhD in a related field—and are affiliated with a university or other respected institution.
- Cite multiple peer-reviewed studies. Use PubMed or Google Scholar to check (at least the abstract) whether a study's findings are consistent with how the research was represented to you. Look for human studies; beware of the use of animal or cell culture studies to prove a point.

Janie Oyakawa eventually reversed her decision on vaccines, but it took both an emotional connection and critical thinking for her to reassess her beliefs. She developed an online friendship with a tattooed, baby-wearing mama who tended backyard chickens, but then she discovered that her new friend was also a vaccine researcher at Johns Hopkins. That gave her an image—a story—of a proud vaccinating mom that contradicted the others in her head. Then she attended a vaccine lecture by a chiropractor and saw how he cultivated fear in his audience. Janie wasn't fooled. "The latent science major" in her started finding holes in his argument, and she knew it was time to reassess her decision on vaccines.[65]

What Janie could finally appreciate was that science gives her the tools to cut through the emotions and stories and to evaluate the evidence for what it really is. Science uses rigorous methods to test vaccines and investigate concerns that arise. We ask researchers to invest years of advanced training in infectious diseases, immunology, and epidemiology, and even

* Are current.
* Don't rely on anecdotes, no matter how compelling.
* Aren't selling products like supplements, e-books, or expensive e-courses claiming to unlock the secrets to good health.
* Don't promote conspiracy theories or ideas completely different from the scientific consensus. (If so, ask yourself, If this source is correct, how many thousands of smart people must be wrong? How likely is this?)

And finally . . .

* Don't assume that a website with a neutral-sounding name offers unbiased information. For example, the National Vaccine Information Center (NVIC) is notorious for spreading anti-vaccine messages and inaccurate information.
* If you're not sure about a source, consider bringing the information you've found there to your pediatrician. He or she can help you assess its accuracy.

then, other scientists check their work and test it again. They do study after study to ensure vaccine safety and efficacy. When Janie focused on the science, she found very clear evidence that the benefits of vaccinating far outweigh the risks, and she felt confident getting her kids caught up on their vaccinations.

For me, choosing to vaccinate my daughter is about a few things. It's a tiny tribute to Frankie and his short life. It's also because I know that, despite reading hundreds of scientific papers on vaccines, I'm not a vaccine expert. The experts in this field have devoted their lives to the science of disease prevention, and they're also parents and grandparents and aunts and uncles. They go to work every day to improve the health of children, and I trust their knowledge on vaccines. I will not look to celebrities or conspiracy theorists for advice about the health of my child. Most of all, my choice to vaccinate is about critical thinking and evaluation of the scientific evidence. Among all of the topics I cover in this book, the choice to vaccinate on schedule is backed by the strongest, clearest, biggest pile of evidence. It's one of the best things you can do to protect the health of your child and others in your community.

✲ RECOMMENDED READING ON VACCINES

I have only one chapter in this book to discuss vaccines, but there is so much more to learn. You can spend years trying to navigate around the Internet looking for reliable information on vaccines and decades reading the scientific literature. I've found it most helpful to sit down with a good book that explores the science, politics, and history of vaccines in detail. Here are some of my favorites:

Deadly Choices: How the Anti-Vaccine Movement Threatens Us All by Paul A. Offit (2010)
Do Vaccines Cause That?! A Guide for Evaluating Vaccine Safety Concerns by Martin G. Myers and Diego Pineda (2008)
The Panic Virus: A True Story of Medicine, Science, and Fear by Seth Mnookin (2011)
Vaccine: The Controversial Story of Medicine's Greatest Lifesaver by Arthur Allen (2007)
Vaccines and Your Child: Separating Fact from Fiction by Paul A. Offit and Charlotte A. Moser (2011)
Your Baby's Best Shot by Stacy Mintzer Herlihy and E. Allison Hagood (2012)

Here are some reliable Internet sources:

Centers for Disease Control and Prevention (CDC), Vaccines and Immunizations page, http://www.cdc.gov/vaccines
The History of Vaccines, www.historyofvaccines.org
Institute of Medicine, Reports (search for topic of interest), http://www.iom.edu/Reports.aspx
Institute for Vaccine Safety at Johns Hopkins Bloomberg School of Public Health, www.vaccinesafety.edu
Vaccine Education Center of the Children's Hospital of Philadelphia, http://www.chop.edu/service/vaccine-education-center/home.html

9 ⁂ GETTING STARTED WITH SOLID FOODS

When and How to Begin

There is nothing quite like a new baby to make the older generation reflect on how things have changed since they were parents. After Cee was born, my mother-in-law unearthed boxes of keepsakes from my husband's baby days. Among the most interesting artifacts was his 1975 baby book. My favorite page is a carefully recorded list of early foods, titled "Growing Appetite." It begins with a mix of breastfeeding and formula but quickly adds rice cereal at 1 month, applesauce at 2 months, meat at 4 months, and egg yolk at 5 months. This was the precise protocol recommended by the pediatrician, and my husband's parents followed it faithfully. From about the 1950s through the 1990s, most infants in the United States, Canada, and Europe were fed in a similar way.[1]

But as we approached the transition to solid foods with Cee, nobody offered us a protocol, and the advice we were given was conflicting. Her pediatrician recommended starting solids between 4 and 6 months, but breastfeeding websites and books urged us to wait until at least 6 months. Some recommended infant cereal and pureed fruits and vegetables, and others extolled the benefits of an approach called baby-led weaning, in which babies entirely feed themselves. Like many new parents, I was confused, and my fancy PhD in nutrition wasn't much help.

I fretted so much about the transition to solids that I totally missed the joy of it. In hindsight, I realize that my worry led us to start off on the wrong foot with feeding. And feeding is at the heart of our relationships with our babies. It starts with how we breastfeed or bottle-feed, with respect for our baby's cues of hunger and satiety, and as we start solids, that back-and-forth conversation about feeding continues to be vital. It can't be based on a protocol or prescribed timeline. Yet, so much of the advice we receive about starting solids implies there's a way that works for every baby: what foods, in what form, and at what age. Starting solids is not just another milestone; it's the beginning of a lifetime of happy and healthy eating for your child.

In this chapter and the next, I tackle many of the questions that I had about feeding as a new mom. I offer you not protocols but knowledge to set the foundation of your confidence. This chapter focuses on when and how to start solid foods, and the next chapter explores what foods to feed.

WHY DO BABIES NEED SOLID FOODS?

Babies go through some incredible nutritional transitions in their first year of life. Before birth, their growth and development is almost entirely fueled by glucose, a simple sugar, which crosses the placenta from mom's blood to baby's. At birth, a newborn has to rapidly adapt to getting nutrients from milk—high in fat and lactose—through his inexperienced digestive tract. And then, when you begin adding solid foods to his diet, he has to make the transition to digesting a wide variety of complex foods. You may start off feeding purees and mushy cereals, but by the end of his first year, your baby should be eating many of the same foods as the rest of the family. Scientists call this transition to solid foods "complementary feeding." Ideally, it takes place against a backdrop of continued breastfeeding, so the goal is to include solid foods that complement the nutrition provided by breast milk. (If your baby is predominantly formula-fed, do not fear; I'll be addressing considerations for complementary feeding with formula throughout the chapter.)

If we want to understand why and when our babies need complementary foods, we must begin with an understanding of breast milk. Breast milk is the evolutionary first food for babies; it is close to nutritionally complete for about the first six months of life (vitamins K and D being notable exceptions). But after about six months, several nutrients become a concern for the exclusively breastfed baby. The first of these—the nutrient most likely to become deficient in a diet of breast milk alone—is iron. I warn you that I'm going to talk about iron a lot in this and the next chapter, because it's sort of a nutritional bottleneck that affects when babies need solids and what types of foods are best.

As I discussed in chapter 2, babies are born with a certain endowment of iron, passed from mom to baby during pregnancy. Babies need iron to support their rapid growth and development, and because breast milk is low in iron, most of a young infant's daily iron needs are met by slowly drawing down his iron stores. For most, the iron endowment is depleted around 6 months of age, but depending in part on when the cord was clamped, this can range from 3 to 8 months.[2]

The picture is a little different for formula-fed babies. Formula is heavily fortified with iron, so it provides more than enough of the nutrient regardless of a baby's solid food intake. This is why breastfed babies are at greater risk for developing iron deficiency and even anemia during late infancy, particularly if they are slow to start solid foods.[3] This should not be interpreted as a shortcoming of breast milk but rather as an indicator that it is important and natural for babies to begin eating solids by around 6 months, when their iron stores run low. Iron deficiency, particularly when it is accompanied by anemia, is associated with long-lasting cognitive and behavioral deficits (see chapter 2). Appendix G provides a detailed explanation of how we calculate babies' dietary iron requirement.

Iron isn't the only nutrient of concern for older breastfed infants. Zinc, a mineral important for normal brain development, growth, and immune function, is also limiting. The zinc concentration in breast milk declines sharply over the first several months of life, and by around 6 months, it isn't enough to meet the baby's needs.[4] For both iron and zinc, giving mom a supplement doesn't improve breast milk levels.[5] Even in developed countries like Sweden and the United States, as many as 20% to 40% of 12-month-olds may have low iron or zinc, with breastfed babies at greatest risk.[6]

The period of 6 to 12 months is recognized as one of the most nutritionally vulnerable times in childhood.[7] For their size, babies of this age have some of the highest nutrient needs of the lifespan, necessary to fuel growth and development. For example, relative to the calories that they're consuming, 6- to 12-month-old infants need nine times as much iron and four times as much zinc as an adult male.[8] Besides iron and zinc, other

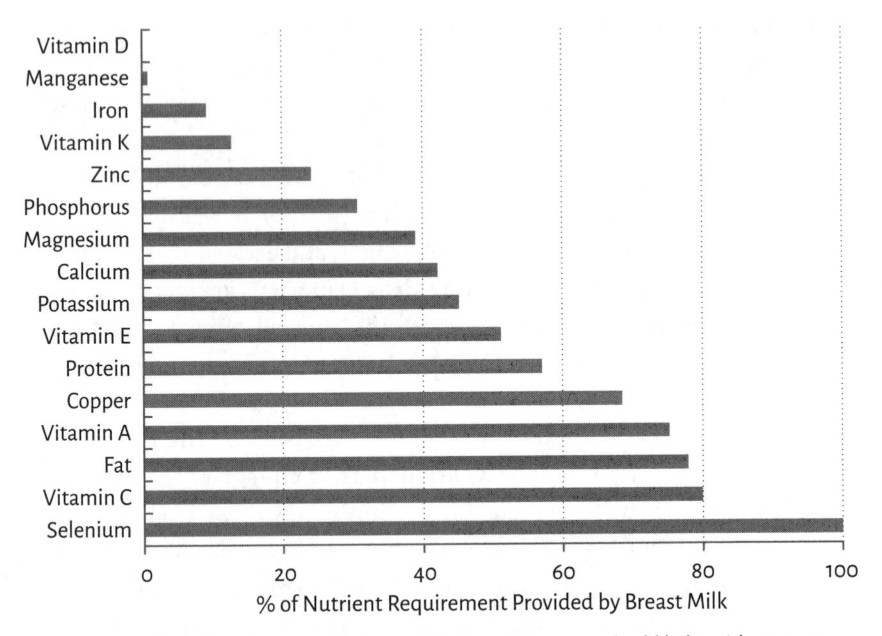

Percentage of nutrients provided by breast milk for a 6- to 8-month-old baby with average breast milk intake (600 ml/day). *Sources:* Breast milk data and most nutrient requirements from the World Health Organization, and all other data from the Institute of Medicine, in C. M. Chaparro and K. G. Dewey, "Use of Lipid-Based Nutrient Supplements (LNS) to Improve the Nutrient Adequacy of General Food Distribution Rations for Vulnerable Sub-Groups in Emergency Settings," *Maternal and Child Nutrition* 6, suppl. 1 (2010): 1-69.

nutrients can also become deficient in a diet of breast milk alone after about 6 months of age. Breast milk is an incredibly nutritious food, and it's amazing in lots of ways, but it isn't a complete diet for the older infant.

So, solid foods are important. And yet, their importance is often downplayed. Breastfeeding advocates love to repeat the catchy phrase "Food before one is just for fun!" The spirit of this comes from a good place: to reassure moms who may be feeling undue pressure that their breast milk isn't enough and that it is time for their babies to get some "real" food. But many moms have instead interpreted this message to mean that breast milk provides all the nutrients that their baby needs for the first year of life. The World Health Organization tirelessly promotes breastfeeding around the world, but it is also unequivocal about the necessity of solid foods: "Complementary feeding should be *timely*, meaning that all infants should start receiving foods in addition to breast milk from six months onwards. It should be *adequate*, meaning that the complementary foods should be given in amounts, frequency, consistency and using a variety of foods to cover the nutritional needs of the growing child while maintaining breastfeeding."[9]

This is not to say that feeding can't be fun, and it isn't meant to scare you. It isn't that hard to meet the nutrient requirements of a growing baby if you offer a variety of nutrient-dense foods, which I explore in detail in the next chapter. Most babies at this age love trying new foods, and it's fun to watch them explore.

If your baby is mostly or completely formula-fed, then you don't have to worry much about the nutrient bottlenecks discussed above. Formula is designed to meet the nutrient requirements of babies through the entire first year, and iron and zinc deficiency are rare among formula-fed infants, so long as their formula is properly fortified with both nutrients.[10] However, it is still important to start solids in a timely manner. As you'll read in the following sections, there are long-term health issues associated with introducing solids too early or too late. And besides, this is an important time of learning for your baby, a chance to explore lots of tastes and textures. So, if you're feeding formula, you can relax a bit about nutrition, but the rest of this information fully applies.

WHEN TO START SOLID FOODS: WHAT IS THE OFFICIAL ADVICE?

The official advice on starting solid foods depends on which official you ask. The AAP says that most babies are ready to start solids between 4 and 6 months of age, and many European pediatric and food safety bodies agree.[11] But even with these guidelines, advice from pediatricians isn't consistent, and families make lots of different choices. A 2005-7 survey of U.S. moms found that 40% introduced solids before 4 months, and more than half of these said they did so because "a doctor or other health care professional said my baby should begin eating solid food."[12] On the other hand, another U.S. study found that 19% of moms waited until after their baby was 7 months old to introduce solids.[13]

Advice about solid foods is also intertwined with breastfeeding advice, and this is where it really starts to get confusing. Since 2001, the WHO has recommended that babies be exclusively breastfed for the first six months of life, meaning that they shouldn't consume any formula or solid foods until 6 months of age.[14] The AAP's Section on Breastfeeding and the U.K. Department of Health second this opinion.[15] (Yes, you read that right. The AAP is divided on this question, depending on whether you ask the nutrition or the breastfeeding experts.) But this is really a recommendation for the optimal duration of exclusive breastfeeding, not the optimal time to introduce solid foods. The two are obviously closely related, but they aren't one and the same, particularly for babies that aren't exclusively breastfed.

Most everyone agrees that infants shouldn't have solids before 4 months of age, but nutrition and public health experts continue to argue the question of whether to start solids between 4 and 6 months or *at* 6 months. It may seem like an academic debate, but to moms and dads trying to make the best choices for their babies, it has real-life implications.

Take college biology professor and single mom Valerie Wheat. Valerie exclusively breastfed her daughter, Caitlin Jo, from birth. But after four months, Valerie was not only going back to work but also starting a new job, and Caitlin refused to take a bottle. "She wanted nothing to do with a synthetic nipple," Valerie told me, and yes, she tried lots of different tricks and types of nipples. "She just screamed."[16]

Valerie's mom was caring for Caitlin Jo during the day, and it wasn't going well. Caitlin was hungry and let everyone know it. They started thinking about trying some solid foods. Valerie wrote to me: "My plan was to exclusively breastfeed for 6 months. I really wanted to wait even later than that, as I'd read on the Internet that this was a minimum goal . . . But here I was trying to start a new job with a baby that would scream for hours while I was gone." Valerie's mom started giving Caitlin some vegetable purees, which helped. Caitlin continued to breastfeed on demand when Valerie was home, and according to Valerie, her daughter is now a "strong and smart and amazing" 4-year-old. But this decision to start solids before 6 months caused Valerie a lot of anxiety and made her transition back to work more difficult.

A mom like Valerie, weighing the pros and cons of offering solid foods, might want to know more about the studies behind the recommendation for six months of exclusive breastfeeding. She might be interested in knowing that some researchers question its applicability to developed countries.[17] As the controversy on this topic suggests, there is complexity and nuance that can't be conveyed in a simple public health recommendation. In the following sections, we'll look at how the timing of introducing solids might affect your baby's health in the short and long term, so that you can make your own decision.

STARTING SOLIDS: IMPACT ON IMMEDIATE HEALTH

To my knowledge, only three randomized controlled trials (RCTs), the highest quality study design, have investigated the optimal timing for the introduction of solid foods. Two were conducted in the 1990s in Honduras.[18] In these studies, one group of mothers was asked to exclusively breastfeed their babies until 6 months, and the other was asked to start solids at 4 months and was provided with jarred commercial foods and advice on sanitary feeding practices, reducing the risk of foodborne illness. There were no obvious differences in growth between the two groups of babies, suggesting that both ways of feeding provide enough calories and nutrients to support normal growth. However, the six-month exclusively breastfeeding babies had lower hemoglobin and iron stores at 6 months. The third RCT, conducted in Iceland with results published in 2012 and 2013, found similar results: no difference in growth but lower iron stores in babies exclusively breastfed for six months.[19]

Observational studies back these findings. A 2012 Cochrane review, sorting through 23 studies (11 from developing and 12 from developed countries), concluded that exclusive breastfeeding until 6 months of age usually provides adequate nutrients for growth.[20] However, the authors are cautious in their conclusions: "The data are insufficient to rule out a modest increase in risk of under nutrition with exclusive breastfeeding for six months and grossly inadequate to reach conclusions about the effects of prolonged (more than six months) exclusive breastfeeding."[21]

What about risk of illness? In developing countries, six months of exclusive breastfeeding (compared with four months) decreases a baby's risk of gastrointestinal (i.e., diarrheal) and respiratory infections.[22] In some settings, the increased risk of diarrhea associated with starting solid foods is dramatic—as much as 13-fold in the rural Philippines, for example.[23] This is probably the most important factor driving the WHO's decision to encourage exclusive breastfeeding for six months. In parts of the world where lack of access to clean water or refrigeration means that feeding solids puts babies at greater risk of exposure to infectious pathogens, the WHO recommendation makes good sense.

In industrialized countries, however, the risk-benefit calculation is a bit different. Among the 12 studies conducted in developed countries, the 2012 Cochrane review found only one in which starting solids earlier increased the risk of illness. This large observational study found that babies exclusively breastfed to 6 months old had a decreased chance of a gastrointestinal infection compared with those exclusively breastfed until just 3 to 4 months (with continued breastfeeding but introduction of some formula and/or solids).[24] The difference equated to about one more case of diarrhea for every 42 babies starting solids early. However, there was no difference in the incidence of hospitalization for gastrointestinal infections, so the increased risk appeared to be in minor illnesses. For all we know, those extra infections could have been avoided with more careful food preparation and storage. And this analysis includes some babies starting solids as early as 3 months, so we don't know how it might apply to those starting at 5 months, for example. Meanwhile, the study found no difference in other health outcomes, including respiratory or ear infections, between babies starting solids early or late. The Cochrane review includes 100 pages of analyses of all sorts of outcomes, based on many studies, but there were no other differences in the health of babies exclusively breastfed until 3 to 4 months versus 6 months.[25]

Other studies provide more reassurance that an earlier start at solids probably doesn't make babies sick. Notably, in the Honduran RCTs, where moms were counseled to use sanitary feeding practices, there were no differences in diarrhea or other symptoms, including fever, cough, or nasal congestion.[26] In addition, a longitudinal study of 16,000 U.K. infants found that timing of the introduction of solids for babies—whether breastfed or formula-fed—had no impact on hospitalization for illness.[27]

SOLID FOODS AND CHRONIC IMMUNE DISEASES

Timing of the introduction of solid foods may also affect your baby's long-term health, including the risk of celiac disease, type 1 diabetes, and atopic diseases (i.e., allergies, eczema, asthma, and hay fever). These are all diseases of the immune system. Celiac disease is an immune response to gluten, a protein found in wheat, barley, and rye; it causes inflammation and damage to the small intestine. Type 1 diabetes is caused by a child's own immune system attacking and destroying the cells of the pancreas that produce insulin, which normally helps to regulate blood glucose. In the case of food allergies, the body's immune system attacks food proteins as if they were pathogenic invaders, triggering an allergic reaction. In all cases, the immune system fails to tolerate something that is a normal part of life for most of us, and these are chronic diseases that affect people for the rest of their lives.

When a baby begins eating solid foods for the first time, new proteins bombard his gut. There seems to be an optimal window—a sweet spot—for the immune system to learn about these proteins. Introduce too early, and it appears that the gut and immune system are not yet equipped to respond appropriately. But introduce too late, and the window may have closed; the immune system might only be able to react to the protein, instead of learning to tolerate it. Introducing gluten to a baby before 3 months of age or after 6 to 7 months seems to increase the risk of celiac disease.[28] Similarly, introduction to cereals, both rice and gluten-containing grains, before 4 months or after 6 to 7 months has been associated with a greater risk of type 1 diabetes.[29] As I said, there's a sweet spot.

The story is similar for food allergies: risk appears to be lowest when foods are introduced between about 4 and 6 months.[30] One study found that children first exposed to wheat after 6 months had a fourfold increased risk of a wheat allergy.[31] Another found that children who first had cooked egg at 4 to 6 months had the lowest incidence of egg allergy, whereas those starting egg at 10 to 12 months had a sixfold increased risk.[32] In a third study, researchers observed that Jewish children in the United Kingdom had a 10-fold higher incidence of peanut allergy than Israeli children.[33] Peanut foods were generally avoided in late infancy in the United Kingdom, whereas they were a common part of the diet of the Israeli babies. The hypothesis that exposure to small amounts of peanut foods early in life may help prevent peanut allergy is currently being tested in an RCT in the United Kingdom.[34]

These studies on allergies and introduction to solid foods have emerged only in the past decade. Meanwhile, the prevalence of food allergies is increasing. The CDC reports that food allergies in U.S. children increased by 50% between 1997 and 2011,[35] and peanut allergies are estimated to have doubled between 1997 and 2002.[36] During this same period, officials recommended delaying the introduction of common allergens until the first birthday and beyond, particularly for kids at high risk for food allergies (i.e., those with a parent or sibling affected). That strategy wasn't based on any evidence, and it wasn't effective. In fact, avoidance may even have contributed to the rise in food allergies.[37] The AAP revised its stance on this question in 2008, stating that there was no evidence that waiting beyond 4 to 6 months to introduce common allergens protects children from the development of food allergies.[38] European, Australian, and Canadian pediatric officials agree.[39]

Some studies have suggested that breastfeeding might protect children from allergies, atopic disease, and celiac disease, but the data are mixed on this point. Many studies support a protective effect of breastfeeding on the early incidence of wheezing, asthma, and eczema, but not on food allergy.[40] However, others find no effect and even an increased risk associated with breastfeeding.[41] Some suggest that the conflicting results are due to a failure to account for breastfeeding *at the time of solids introduction* and that this might be a critical factor important to a protective effect,[42] but this hypothesis needs further investigation. This is a tricky area of research because so many factors go into our decisions about how to feed our children, and these could confound the data. For example, a mom who has a family history of autoimmune disease or allergies might be more likely to breastfeed for longer and to delay the introduction of solids for this very reason. I think the jury is still very much out on how breastfeeding may or may not affect the risk of these immune diseases. All we can do is make the best decisions we can with the evidence we currently have.

WHEN TO START SOLIDS: YOUR CHOICE, IN THE REAL WORLD

As with any decision in parenting, there are risks and benefits to starting solids earlier or later. When I look at this research and think about Valerie, who worried about starting her hungry baby on solids before 6 months, I wish I could go back in time and reassure her that she was doing a great job nourishing her baby. I think that the recommendation to start solids between 4 and 6 months is more appropriate, given the data we have. And I think that the recommendation to exclusively breastfeed for six months should be considered a ballpark suggestion, not a benchmark for success or failure.

⁎ PREVENTING AND DETECTING FOOD ALLERGIES

- Breastfeeding may help protect your baby from food allergies, particularly continuing breastfeeding as you introduce solids.
- If your baby is at high risk for allergies and you're feeding formula, there may be a benefit to feeding a partially or extensively hydrolyzed formula for the first six months.
- Start offering solids to your baby between 4 and 6 months of age, beginning with foods that are rarely allergenic, like rice, barley, oats, meat, fruits, and vegetables.
- Gradually introduce potentially allergenic foods, including wheat, eggs, peanuts, tree nuts, fish, and shellfish. Introduce dairy, but avoid feeding cow's milk until 1 year of age (it can increase the risk of iron deficiency).
- Introduce new foods one at a time, and watch for a possible allergic reaction, which usually occurs within minutes to hours. Symptoms include swollen and itchy eyes or mouth, nasal congestion, sneezing or wheezing, diarrhea, vomiting, or hives.
- If your baby has a sibling with food allergies (particularly for peanuts, because this reaction can be so dangerous), or if he appears to have an allergy himself, consult your pediatrician or an allergist to make a personalized plan for introducing foods.

K. M. Järvinen and D. M. Fleischer, "Can We Prevent Food Allergy by Manipulating the Timing of Food Exposure?" *Immunology and Allergy Clinics of North America* 32 (2012): 51-65.

And what about babies who aren't exclusively breastfed? In 2010, just 38% of U.S. infants were still exclusively breastfed at 3 months, and only 16% continued on to 6 months.[43] Most babies are already getting some formula by the time their parents start thinking about introducing solids. Encouraging breastfeeding is a worthy goal, but formula-feeding parents need evidence-based advice too. Feeding formula changes the risk-benefit calculation for starting solids. The chance of contamination or microbial growth is already there if you're making up bottles, but you can minimize it with proper technique. Adding in some other foods is unlikely to be a risk,[44] although you still don't want to start before 4 months. For formula-fed babies, the 4 to 6 month guideline also makes good sense.

As you make your own decision, there are some practical factors that you might also consider. By 4 months of age, breastfeeding or bottle-feeding is usually pretty simple. It's quick and easy and can be done anytime or anywhere with little mess or fuss. In my experience, starting solid foods was messy and slow for the first few months, until Cee started eating more finger foods. (Then the mess was saved by our dog, who took his floor clean-up duties very seriously.) Waiting until closer to 6 months means that your baby can do more feeding himself and join you at the table to share family foods sooner.

I've talked about lots of factors that go into the decision about when to start solid foods, but I have yet to discuss the most important one: your baby. In the next section, I finally take the focus off the calendar and put it where it belongs, with the little one actually doing the eating.

* WEIGHING THE EVIDENCE FOR AGE-BASED SOLID FOODS RECOMMENDATIONS IN DEVELOPED COUNTRIES

Advantages of starting solids between 4 and 6 months of age:

* Lower risk of iron deficiency
* Possible lower risk of food allergies, celiac disease, and type 1 diabetes

Advantages of starting solids at 6 months (it is not recommended to wait much beyond this):

* Possible lower risk of diarrhea
* Baby more developmentally ready for self-feeding, making feeding easier

WANT TO KNOW WHEN TO START SOLIDS? ASK YOUR BABY.

Let's not forget that while the WHO, AAP, scientists, breastfeeding advocates, and parents all across the Internet are arguing about the optimal age to start solids, there are real babies involved. They don't become instantly ready for solids the moment the clock strikes midnight on their 6-month birthday. Rather, they're all developing at different rates, with slightly different nutritional needs and preferences for tastes and textures.

This point was well made by British pediatrician Martin Ward Platt, in a 2009 editorial on "weaning," the British term for starting solid foods (while continuing breastfeeding or bottle feeding). He wrote: "The weaning debate has been largely predicated on the notion that there is some magic age at which, or from which, it is in some sense 'safe' or 'optimal' to introduce solids. Yet it is highly counterintuitive that such an age exists. In what other area of developmental biology is there any such rigid age threshold for anything? We all recognize that age thresholds are legal inventions to create workable rules and definitions, and have no meaning in physiology or development, yet when we talk about weaning we seem to forget this."[45]

Given all the complex factors that I've discussed, a baby's unique development is probably the most important factor of all, and yet it is too often ignored. And because developmental readiness is so variable, identifying it is really up to you.

Developmental readiness for solids is so important because it signals that your baby is ready to be an active participant in the feeding process. He needs to have the oral-motor development necessary to choose to open his mouth, close it over food, and move the food toward the back of his mouth to swallow it without gagging. He also needs to be able to tell you—in his own way—when he's ready for a bite and when he's done. This kind of back-and-forth communication is critical for getting off on the right foot with feeding.

What developmental milestones signal readiness to start solids? First, a baby needs to have the head, neck, and back control to sit upright. This happens in most babies between 4 and 6 months. Babies born prematurely often approach developmental milestones at their corrected age and sometimes on their own timetable altogether, so talk with your pediatrician about developmental readiness if your baby was premature.

To understand the importance of your baby being upright for feeding, feeding expert Ellyn Satter recommends that *you* try eating a bowl of yogurt with your chin down on your chest or looking to the side.[46] It feels funny to us adults, and we have the benefit of already being skilled at maneuvering food in our mouths and knowing how to swallow. Babies are learning all of this by trial and error, and starting solids before they can comfortably sit upright makes their job more difficult and less fun. Besides, if your baby can't sit upright, you can't look one another in the eye while feeding solid foods. How will your baby tell you that he's ready for another bite, or that he's done? Face-to-face interaction is essential.

At about the same time that a baby begins sitting up he often shows interest in watching other people eat and starts grabbing for foods. This is normal development, not an indication that he is necessarily hungry. He's probably most interested in watching you and in exploring objects in his world using his mouth. But if he's 4 months or older, you might begin to try some solids.

When you do start solids, watch your baby carefully to see how he responds to your offer of a spoon of food. If he's ready, he'll lean forward and open his mouth, and he'll accept a bite without pushing it right back out with his tongue (although he may still make a big mess!), showing that he's lost the tongue thrust reflex present in younger babies. If he can do all of these things and then opens his mouth for another bite after swallowing the first, he's probably developmentally ready to eat solids.[47] If he struggles with any of these steps or just doesn't seem to be interested or enjoying the process, that's a good signal to put the food away and try again in a few weeks.

As your baby gets stronger, he'll be more interested in feeding himself. He'll reach for foods in your hand and rake foods forward on a tray to experiment with picking them up. He'll point at foods that he'd like to try and push them away when he's all done. As he develops, you'll find that he quickly outgrows purees, and he may prefer to feed himself if you let him. You can progress to offering lumpier textures and chunks of soft finger foods. Most babies are able to grasp foods and self-feed by 7 to 8 months, if not sooner.[48] The more you let your baby do for himself, the more confident you both will be in his ability to lead the way with feeding.

This is a time of rapid development, and babies are open to tasting lots of foods at this age, so try not to get into a dull cereal and puree rut.

* SIGNS THAT YOUR BABY IS READY TO TRY SOLID FOODS

- Your baby is at least 4 months old.
- He is able to sit up and hold his head up to face you.
- He shows interest in your food.
- If you offer him a bite, he leans forward and opens his mouth with interest.
- He is able to close his lips over a spoon, scrape the food into his mouth, and move most of it from the front to the back of his mouth to swallow. Or, he is able to self-feed soft finger foods.
- He is able to signal to you when he's had enough by closing his mouth or turning his face away. (Respect this!)

A careful study of babies' chewing efficiency showed that the most rapid development of oral skills with solid textures occurred between 6 and 10 months, although babies continued to improve their chewing skills for several years.[49] Babies who eat foods that require chewing also have higher nutrient intakes than those fed purees, probably because finger foods have less water and so provide more nutrients in every bite.[50] Furthermore, among a large, longitudinal U.K. cohort of babies, those that started eating lumpy solids at 10 months or later were more likely to have feeding difficulties at 15 months and 7 years of age.[51]

If your baby is slow to make the transition from purees to finger foods, the important thing is to keep giving him low-pressure opportunities to learn. Offer some finger foods but also provide enough softer foods to satisfy his appetite. He'll explore when he is ready. If your baby doesn't seem to be showing signs of developmental readiness or interest in starting solids—whether pureed or chunky—and he's older than 6 or 7 months, please do talk to your child's pediatrician. He or she may want to assess growth and development and check your baby's iron status. You don't want to push your baby to eat before he's ready, and you don't want any of your own anxiety about feeding to spill into your interactions with your baby. Knowing that he's doing okay in his development and nutrition can help you relax about the feeding process.

BE RESPONSIVE IN YOUR FEEDING

Back when Cee was a baby, I studied the data on age-based recommendations and noticed that she was sitting up and reaching for food, so I started offering her some solids at around 5 months of age. I dutifully mixed rice cereal with breast milk and offered Cee a bite. She moved it around in her mouth and then burst into tears. Determined that the problem was just the pasty cereal, I pureed some sweet potatoes and mashed some avocado. Cee loved watching me make these colorful foods, but again, tasting them only led to more distress. I kept at it, thinking that she just needed more practice. But whatever I put on Cee's spoon, and whenever I gently cajoled her to open up her mouth to try a bite, she just wasn't that interested.

After about six weeks of half-hearted dabbling in pureed food, my little family went on a vacation in Hawaii. Determined to relax, I decided that offering Cee her customary spoonful of solids (almost always rejected) every day was too much trouble. Neither of us was enjoying the feeding process much anyway. But then a funny thing happened. One day, I was sitting on the beach eating a banana, and Cee reached out for it and grabbed a piece. She ate more banana in one sitting than all the solids, combined, in her life so far. At a restaurant in the evening, she reached for a soft roll and devoured it. That week on vacation was a breakthrough for us. Cee discovered the joy of eating, and I learned that she really wanted to feed herself. I should have picked up on those signals much earlier, but Cee caught up in no time. Back home, she was ready to try anything that I could present to her in a reasonably soft form. By 8 months, she was eating three square meals a day, with a voracious and adventurous appetite.

All babies are different. Some love being fed purees, and others prefer to feed themselves. Cee was a baby who preferred to feed herself. But as I studied the research on feeding, I wondered about another possibility: perhaps my pressure, my insistence that she try a tiny taste of food each day, even when she wasn't interested, had put a bad taste in her mouth about spoon-feeding. I wasn't being responsive in my feeding. I was paying more attention to the research and rules about starting solids than to my baby, who was trying to tell me that she wasn't quite ready.

My mistake is repeated by new parents every day. And it isn't really our fault. We are given so much advice on solid foods that focuses on ages and milestones that we forget to pay attention to our babies. And just as important as how and when we begin complementary feeding is how we proceed. From the first bite of food and continuing as your baby grows, being responsive in your feeding is most important. Just as with milk feeding, don't push him to eat if he isn't ready or isn't hungry, and don't expect him to finish some arbitrary portion of food if he's signaling to you that he's full. Be sure that he is an active participant every step of the way.[52] Nobody knows better than your baby when he is full. Babies are really good at self-regulation. If you feed a baby more often or more calorie-dense foods, he'll eat less at each meal.[53] This ability to self-regulate appears to decline in the toddler years, probably because parents have convinced their children that their signals of hunger and satiety aren't that important. That's the message a child receives if he declares that he's all done (in whatever language he's using at his age), and the parent says, "How about one more yummy bite?"

Research clearly shows that pressure tactics and feeding don't mix. Pressure comes in many forms. It can mean prompting, like the example just given, or praise ("What a good boy for finishing your broccoli!"), rewards for eating, or coercion, like sneaking a bite of food into your baby's mouth while he's distracted. Over time, pressure during feeding seems to reduce a child's ability to self-regulate food intake and often increases pickiness.[54]

In these days of rising rates of childhood obesity, your early feeding practices can support your baby in healthy growth. A fascinating study found that when mothers were responsive in their feeding of solids, their babies seemed to self-correct their growth trajectories in late infancy.[55] Babies that had gained weight quickly in early infancy showed slower growth between 6 and 12 months, and those that were growing slowly began to gain more. On the other hand, babies whose mothers were more controlling in feeding did the opposite: those that were fatter at 6 months continued getting fatter, and those that were smaller at 6 months stayed small. When it comes to how much to eat, baby knows best.

To feed responsively, begin by creating a supportive environment, giving your baby a comfortable place to sit, facing you, with few distractions, so that both of you can focus on the task at hand. Bringing your baby to the family table where everyone else is enjoying food is usually the best way to create a supportive environment. You offer food that is developmentally appropriate and nutritious. Your baby tells you if he's hungry or full, if he'd like to eat faster or take a break. You respond appropriately, and your baby experiences the satisfaction of knowing that you support him.[56]

IS BABY-LED WEANING A BETTER WAY?

There's a trend in infant feeding called baby-led weaning (BLW), in which a baby is expected to fully feed himself solid foods from the start, completely skipping over spoon-fed purees. In BLW, the baby simply joins the family at the table and begins to eat what he can of pieces of whole foods. BLW proponents claim that this approach to infant feeding encourages greater independence and confidence in babies, develops their hand-eye coordination and chewing skills, and decreases picky eating and mealtime battles.[57] But you know me—I want evidence. Is there any evidence that BLW is a better way to approach the transition to solids?

Several studies have shown that babies fed with the BLW approach ate a variety of foods and that their mothers were more relaxed about feeding than those who chose a more conventional route.[58] However, these weren't randomized trials; the moms in these studies had already chosen to follow BLW. Maybe moms who choose BLW are already more relaxed about feeding, making them well suited to this approach. Anecdotally, families who use the BLW approach often swear by it, saying that it makes feeding solids easy and fun. Since Cee wasn't interested in spoon-feeding, we essentially did BLW without knowing that it had a name and a devout following. It worked well for us.

But without randomized controlled trials of BLW, we can't say whether it is appropriate for all babies. However, given that babies develop at different rates, my guess is that it isn't. In a survey asking parents at what age their babies reached out for solid foods, a prerequisite for self-feeding, only 56% were reaching out at 6 months.[59] Six percent still hadn't reached for food by 8 months, and these same babies were also slower to walk and talk. From a nutritional standpoint, if you expected your breastfed baby to feed himself, but he was still incapable of doing so by 8 months, then your baby would be at high risk for nutrient deficiencies. There is no evidence or logical reason to think that a baby whose motor development is more gradual has lower nutritional needs. All babies have tremendous brain growth during this time, and they all need nutrients to support it. In fact, since iron deficiency can cause developmental delays, I would want to be extra sure that my baby was getting enough iron if he was slow to meet developmental milestones. The authors of the developmental study discussed above concluded the following about BLW: "Our data would

PREVENTING AND RESPONDING TO CHOKING

In the United States each year, more than 12,000 children end up in the emergency room for food choking incidents, and 62% are under 4 years old. Take these steps to prevent choking, and be prepared to respond:

* Your baby should be sitting upright and always supervised during mealtimes. This means no snacking in the car or while running or playing.
* The following are common choking hazards: nuts and seeds, hot dogs, chunks of meat or cheese, whole grapes, hard candy or gum, popcorn, globs of peanut butter, and other round and firm foods. Avoid feeding these foods to babies. Cook fruits and vegetables so that they are soft, or finely grate raw produce (like apples and carrots).
* Know the difference between gagging and choking. Gagging is a reflex that helps move food forward in the mouth for more chewing. Gagging is common in babies learning to eat, but it can sound distressing! If your baby is gagging, he should recover quickly and be able to breath and talk. Choking is usually silent, which makes it extra scary—and extra important that children are supervised while eating.
* Know the Heimlich maneuver and CPR for babies, children, and adults. The American Heart Association offers community and online courses (www.heart.org).

M. M. Chapin et al., "Nonfatal Choking on Food among Children 14 Years or Younger in the United States, 2001-2009," *Pediatrics* 132 (2013): 275-81.

suggest that there might be risk in this method for children with relatively slower development . . . However, mild developmental delay is not usually recognized until the second year of life, and even children at the bottom of the normal range could be vulnerable if the method was applied too slavishly. Parents should probably be advised that BLW is only realistic so long as their child is reaching out for and mouthing objects by the age of six months."[60]

One purported benefit of BLW is that it encourages baby's self-regulation of appetite, because self-feeding inherently gives him full control of the pace of eating. A 2012 study found that BLW babies were less likely to be obese as toddlers, but this study revealed a downside as well: BLW babies were also more likely to be underweight, highlighting the risk discussed above.[61] BLW definitely ensures that the baby leads the way with feeding, but you can achieve the same thing by spoon-feeding responsively. Regardless of how you feed, you should be responsive to your baby's signals of hunger and satiety. In this sense, all feeding should be baby-led. The tools and the texture are just details.

Parents and pediatricians often worry that BLW might increase a baby's risk of choking, and parents making the change from spoon-feeding to finger foods might have the same fear. As far as I can tell, there are no data comparing choking accidents in BLW versus spoon-fed babies. However you feed, know how to prevent and respond to choking.

My opinion about BLW is that it adds unnecessary restrictions to feeding solids. Besides limiting when your baby can start solids, it also limits foods to those that your baby can handle and gum up on his own, reducing your choices for types of foods that you can offer. I suggest a more relaxed approach that allows you to start with purees if it seems appropriate for your baby and then advance quickly to new textures, as he is ready. You will help your baby with many things in life. Feeding solids may or may not be one of them, but there's no shame in giving him a hand if he needs it. That's just responding to his needs, which is good parenting.

Of course, there's one big question we haven't even tackled yet: what foods should you give your baby? I devote the entire next chapter to exploring appropriate complementary foods.

10 ⁂ EAT, GROW, AND LEARN

The Best Foods for Babies

Eve, a new mom in Littlehampton, England, was preparing to offer her baby boy solid foods for the first time. Like many new parents, she wasn't quite sure how to start, and she turned to the Internet for advice. She posted her question on the Facebook page of *Natural Mother Magazine*, a busy page with 30,000 followers at the time.[1] Eve wrote: "I am hell-bent on giving him the best start in life, so he will be fed organic and ideally fresh local produce if it kills me." She goes on to say that she has heard that baby rice cereal is "a bad option," but that leaves her at a loss about what to start with instead. "One source said the first food should be egg yolk, and another said raw liver! Please HELP!"

Responses poured in from parents around the world, and the recommendations for first foods read like a holiday feast. Smashed banana, pumpkin, and avocado. Carrots, potatoes, and pears. Pureed lamb and rabbit. Bone and meat broth. Quinoa, amaranth, and oats. Egg yolk scrambled with breast milk and organic butter. And, oh yes, liver—cooked or raw, your choice.

The *Natural Mother* demographic is just one slice of today's parenting population, but their feeding preferences mirror the current shift toward greater emphasis on whole, or unprocessed, foods. Surely that pasty, processed rice cereal isn't an optimal first food for a baby. How unnatural, we think.

Throughout human history, how and what we feed our babies has varied tremendously over time and across cultures. At this time, our most recent traditions—the way our parents fed us—are based on refined infant cereals and pureed fruits and vegetables, but go back another few generations, and you'll find an entirely different set of traditions. Dr. L. Emmett Holt's classic text *Diseases of Infancy and Childhood* was published in 11 editions between 1897 and 1953 and illustrates a radical shift in advice over that period. Early editions recommended feeding beef juice, thin cereal porridges, and custards, and specifically avoiding vegetables until 3 *years* of age. By 1953, Holt recommended introducing a variety of foods, including vegetables, by 5 months.[2] How we've fed infants over the course of time hasn't been rooted in particularly strong scientific evidence but rather in evolving traditions and cultural norms.

Where does this leave us as new parents, trying to feed our babies well? If we try to crowd-source feeding advice, as Eve did, we find vastly different answers depending on the crowd we ask. Who is right? Can we see some evidence, please?

In this chapter, I ignore the trends and traditions and identify some of the best foods for babies based on the science of infant nutrition. To do this, we have to start by taking a step back and looking at the bigger picture. Breast milk or formula will remain a major source of calories and nutrients for your baby, so the solid foods that you choose should be a good complement so that baby's small tummy is full of good stuff—both milk and solids. For a breastfed baby, which foods to feed really depends on the nutritional gaps between what breast milk can provide (see the graph in chapter 9) and what your baby needs. Because formula is fortified with minerals and vitamins, getting these nutrients from solid food is less critical to a predominantly formula-fed baby. However, we still want to offer a wide variety of nutritious foods, because this period is an opportunity for all babies to learn to eat—to experience the tastes, textures, and flavors of a healthful diet.

AN EVOLUTIONARY PERSPECTIVE ON INFANT FEEDING

In the United States and other Western countries, we rely heavily on iron-fortified infant cereal to meet the nutrient needs of older infants. In 2008, nearly 80% of infants aged 6 to 9 months consumed infant cereal daily.[3] But fortified cereal has been around in the United States for only a half-century or so. Some parents don't want to feed it, and some babies, like my Cee, refuse to eat it. (Have you tried it? It is rather metallic-tasting.) How did babies get enough iron before we had fortified cereals? Surely our ancestors and families living in less industrialized societies can teach us a thing or two about feeding babies without processed foods.

Because of the high nutritional needs of older infants and the risk of malnutrition in this age group around the world, child nutrition experts work hard at figuring out how best to feed babies using traditional and locally available foods. But when they look at a diet of on-demand breastfeeding, plus legumes, eggs, fish or chicken, green vegetables, and a staple grain each day, they find that babies manage to get only about 30% to 50% of their daily requirement for iron and fall short on zinc as well.[4] It is only when babies are also given a serving of liver (high in both iron and zinc) every day that they are able to get adequate nutrients. The sad reality is that many babies around the world are fed thin, grain-based gruels and don't have access to fortified foods or supplements, so it's no wonder that nutritional deficiencies and faltering growth are exceedingly common in babies and young toddlers in poor countries.[5] In developed countries like the United States, iron-fortified cereals probably save many babies from deficiencies. According to the WHO, "Average iron intakes of breastfed infants in industrialized countries would fall well short of the recommended intake if iron-fortified products were not available."[6] Indeed, prior to the widespread use of today's fortified cereals, iron deficiency was much more common in young children in the United States.[7] (If you're wondering how we know that babies need so much iron, see appendix G for a detailed explanation of how the iron requirement is calculated.)

This is puzzling, isn't it? How is it that we have this beautiful, complete food for young infants—breast milk—that suddenly becomes nutritionally inadequate in late infancy, and we can't make up the shortfall using traditional foods? Have babies always run the risk of iron deficiency, throughout our evolutionary history?

Dr. Kathryn Dewey, professor of nutrition at the University of California, Davis, has been trying to answer this question. And after studying breastfeeding and complementary feeding for 30 years, she believes that the answer lies in the human history of 10,000 to 15,000 years ago, during a time before agriculture, when we were hunter-gatherers reliant on foods caught or harvested from the wild. As she wrote in a 2013 review paper, wild game, fish, shellfish, and insects dominated the diets of hunter-gatherers, making up 45% to 65% of calories.[8] The rest of the diet came from wild plants, including leaves, flowers, nuts, seeds, roots, and fruits. Overall, this pre-agricultural diet would be high in many of the nutrients that are essential to babies' growth and development. The animal protein sources are rich in iron and zinc, and mothers probably pre-chewed these otherwise tough foods to make the nutrients available to their babies.[9] Baby-sized portions of a hunter-gatherer diet, along with continued breastfeeding, would nicely meet the needs of infants. They fall a little short in iron by 6 to 8 months, but this difference might easily be made up by delayed cord clamping, which would have been the norm at hunter-gatherer births.

The pre-agricultural diet was also probably higher in omega-3 polyunsaturated fatty acids, in part because of the regular consumption of fish, but also because omega-3s were more prevalent throughout the food chain. Many wild plants are relatively high in omega-3s, and these are incorporated into the meat of wild game. As humans domesticated plants for agriculture, seeking higher yields and better stability for storage, there was a gradual shift in our food supply toward more omega-6 fatty acids and less of the more healthful omega-3s.[10] Of particular importance for babies is the omega-3 docosahexaenoic acid (DHA), which is necessary for vision and brain development.[11] DHA is found in breast milk, but its concentration is dependent on the mother's diet, and modern complementary foods often contain little DHA.[12]

Maybe our hunter-gatherer ancestors were onto something when it came to feeding babies, but before you pick up a spear and head out the door to look for some wild game for your baby, or start digging around for insects at your neighborhood playground, let's put this into a modern perspective. We aren't hunter-gatherers. Even if we wanted to be, there isn't enough wild game for all of us to go back to hunting 50% of our food. And it would put too much of a strain on the environment to increase animal agriculture so we could eat much more meat or aquaculture so we could eat more fish. I personally like relying on my local farmers to provide my family with delicious fresh vegetables, and I like beans and whole grains.

However, I do think that taking an evolutionary perspective can help us to make sense of the nutritional bottlenecks in infancy and to make smart choices with all of our modern options. If we're not feeding our babies like hunter-gatherers, then we at least need to make sure we're meeting their nutrient needs with our modern foods.

WHY YOUR BABY SHOULD BE A MEAT EATER

Here's something that hunter-gatherer parents almost certainly got right: they fed their babies meat. Including a daily serving of meat in your baby's diet can help meet both the iron and zinc requirements.[13] One reason that meat is such a valuable food for babies is that it contains heme iron (i.e., iron bound to hemoglobin or myoglobin), and heme iron is absorbed relatively efficiently from the digestive tract into the bloodstream. In these foods, about 25% to 35% of the iron is absorbed for use in the body—a measure called bioavailability.[14] Plant foods such as legumes, grains, and vegetables contain nonheme iron, which has a much lower bioavailability, often 5% or less.[15] However, there are many dietary factors that can increase or decrease the absorption of nonheme iron. (See page 186.)

Not only is the heme iron in meat highly bioavailable, but it also improves the absorption of nonheme iron included in the same meal.[16] Meat is easily digested and makes an ideal first baby food if pureed or slow-cooked. And yet, in 2008, less than 10% of U.S. infants aged 6 to 9 months consumed any type of meat daily.[17] That's too bad, but hopefully this number will increase as more parents and pediatricians understand the benefits of meat. The AAP now recommends meat as one of a baby's first foods.[18]

In addition, meat is a good source of fat—in particular, arachidonic acid, a polyunsaturated fatty acid (PUFA) found only in foods of animal origin. Beef and pork contain 1.3% to 1.6% arachidonic acid, whereas chicken has just 0.3% and breast milk has 0.5%. Arachidonic acid is thought to play a role in infants' psychomotor development, and one study found that infants who ate more meat scored higher in fine and gross motor development.[19] Another found that breastfed babies eating pureed beef had greater head circumference and tended to be more engaged with people and the environment than those consuming a vegetar-

✲ BABY-FRIENDLY SOURCES OF IRON

FOOD	*SERVING SIZE*	*IRON (MG/SERVING)*
	Heme sources (25% to 35% absorbed)	
Home-cooked foods	30 g (~1 oz)	
Chicken liver		3.9
Beef liver		2.0
Beef (ground)		0.9
Sardines and clams (canned)		0.8-0.9
Turkey		0.6
Tuna (canned)		0.4
Chicken		0.3
Salmon		0.2
Commercial baby food—meats	71 g (2.5 oz jar)	0.5-1.0
Whole egg (>90% of iron is in the yolk)	1 (large)	0.9
	Nonheme sources (~5% absorbed)	
Blackstrap molasses	1 Tbsp	3.7
Home-cooked foods	~1/4 cup (weight as below)	
White beans	50 g	1.9
Lentils	50 g	1.7
Other beans	50 g	1.1-1.4
Spinach	50 g	1.8
Oatmeal	20 g (dry)	0.9
Whole wheat bread	30 g (1 slice)	0.8
Quinoa	50 g	0.7
Green peas	40 g	0.6
Rice, pasta, bulgur, barley	50 g	0.3-0.7

continued

Apricots, dried	30 g	0.8
Raisins	40 g	0.8
Fortified cereals		
Infant cereal	15 g (dry)	5.0
Oatmeal, regular instant*	20 g (dry)	5.0

Source: U.S. Department of Agriculture, Agricultural Research Service, "USDA National Nutrient Database for Standard Reference, Release 26," Nutrient Data Laboratory Home Page, 2013, http://www.ars.usda.gov/ba/bhnrc/ndl.

Note: Recommended iron intake for infants depends on food bioavailability. If 10% bioavailability, about 9 mg iron per day is recommended (see appendix G).

*In fortified cereals not made specifically for infants, including cold breakfast cereals, iron is often present in forms that are less bioavailable than in infant cereals.

ian diet.[20] Choosing grass-fed meat, if possible, will also give your baby more brain-building omega-3 fatty acids, including DHA.[21]

Liver is often recommended as a good food for infants, because it is one of the best sources of bioavailable iron and zinc. It has the additional advantages of being relatively inexpensive, easy to prepare, and well liked by many babies. However, liver also accumulates potentially toxic levels of vitamin A and other fat-soluble compounds, so there is at least a theoretical concern about the safety of feeding liver to babies frequently.[22] If your baby enjoys liver, consider limiting it to a few servings per week. Raw liver poses a risk for foodborne illness, so be sure to cook it before serving.

Fish is also a good source of bioavailable iron and zinc, as well as an excellent source of omega-3 fatty acids. Sardines and some shellfish, like mussels and clams, are particularly good sources of iron.

FEEDING YOUR BABY EGGS AND DAIRY

Egg yolk is an excellent first food for babies. It is, after all, a concentrated source of nutrients meant to support the earliest stages of development, albeit for a chick. Egg yolk is high in fat and a good source of iron. Since these nutrients are concentrated in the yolk, and babies just starting solids eat only a small amount each day, the yolk is the most valuable part of the egg. It's easy to separate out the egg yolk before cooking or from a hard-boiled egg, and it's already a nice soft texture for feeding.

There's a little confusion around egg as an iron source. Egg contains both heme and nonheme iron, but studies have shown that whole egg *inhibits* iron absorption. This is probably due to the egg white protein rather than the yolk.[23] A randomized controlled trial (RCT) found that feeding four egg yolks per week between 6 and 12 months (in addition to a well-balanced diet) resulted in a small improvement in iron status.[24] Thus, egg yolk alone seems to be at least a decent source of iron. As your

⁎ OPTIMIZING IRON ABSORPTION

Absorption of nonheme iron, the form found in plant foods, is highly variable and depends on other components present within the same meal. Planning your baby's meals with this in mind can help optimize iron absorption.

The following food components *decrease* the absorption of nonheme iron when included in the same meal:

- Phytates, found in high levels in whole grains, legumes, seeds, and nuts (but you can reduce phytates by soaking, sprouting, or fermenting foods)
- Calcium, including in dairy products
- Soy products
- Phenolic compounds, found in tea, coffee, and chocolate (I've added these just for the sake of completeness, since I doubt you feed these to your baby)

To *increase* nonheme iron absorption, include any of the following in the same meal:

- A good source of vitamin C (at least 25 mg; see the table on page 191).
- A source of heme iron (meat, poultry, fish)

baby gets older and eats greater quantities of other iron-rich foods, you can start feeding whole egg.

Eggs are also a good source of omega-3 fatty acids. In breastfed infants, levels of DHA in red blood cells have been shown to decline between 6 and 12 months, most likely because of a decreasing intake of DHA-rich breast milk. Because our evolutionary diet was probably much higher in omega-3s, it is reasonable to think that we should try to maintain DHA levels during this period of rapid brain development. The RCT mentioned above found that feeding four regular egg yolks per week maintained red blood cell DHA during late infancy, and feeding omega-3–enriched eggs actually increased DHA levels.[25] Fatty fish is also an excellent source of DHA, but because of concerns about mercury, the Food and Drug Administration recommends that pregnant or breastfeeding women and young children consume only two to three servings per week.[26] Egg can be eaten daily without this concern.

Purchasing omega-3–enriched eggs for your baby may be a good choice. These are made by feeding laying hens a diet naturally high in omega-3s (more like what wild birds might eat), usually in the form of flax or fishmeal. These eggs are particularly high in α-linolenic acid, which boosts the total omega-3 count listed on the package but isn't the most beneficial fatty acid on its own. DHA is also increased, and although the amount varies substantially with the hens' diet, omega-3–enriched eggs usually provide at least 100 mg of DHA, more than double that found in a typical supermarket egg.[27] One study found a significant improvement in visual acuity of infants fed 83 mg of DHA from egg yolk each day from 6 to 12 months.[28]

An even better choice may be eggs from pasture-raised hens, which are naturally higher in DHA than eggs from conventional grain-fed hens.[29] The problem here is that your local grocery store may or may not carry eggs from real pasture-raised hens—those that have been able to eat grass, seeds, and insects to their heart's content. In the United States, the label "free range" on poultry products means only that the hens have access to the outdoors, which could just be a strip of concrete.[30] The term "pasture-raised" on the label is currently unregulated, so your best bet is to buy eggs from a local farmer (check your farmers' market) or to raise your own.

If you're worried about your baby getting too much cholesterol from eggs, don't be. In the RCT, feeding babies four egg yolks per week did not affect plasma cholesterol. Besides, breast milk is naturally high in cholesterol, and this is probably a good thing for babies.[31] Egg yolks are also rich in the nutrient choline, which is important for brain development in early childhood.[32]

Between 6 and 12 months, dairy products should play only a small role in your baby's diet. Not only is cow's milk low in iron, but the iron has low bioavailability, and components of the milk inhibit the absorption of iron from other foods. In addition, in babies less than 12 months old, cow's milk can increase blood loss in the stool, depleting their bodies of iron.[33] Wait until your baby is at least 12 months old to begin serving cow's milk as a regular beverage. Dairy foods like full-fat yogurt and cheese are good sources of fat and calcium, but feed them as snacks separate from meals with iron-rich plant foods so they don't get in the way of iron absorption. After 12 months, your baby's iron requirement decreases (because she's not growing as fast as she did in the first year), and she should be eating more iron-rich foods, so serving cow's milk with meals is okay.

BUT I'M A VEGETARIAN!

If you're vegetarian, feeding your baby meat and other animal products may not appeal to you. You might consider including meat in your child's diet for at least the first couple of years, but if not, you really need to take special care to ensure that your baby eats good sources of fat, iron, and zinc. Babies are thought to need at least 25% of their calories from fat to support growth and brain development, and plant-based foods alone can't meet this requirement.[34] Although whole grains and legumes, including soy, contain a reasonable amount of iron, its bioavailability is very low. Including full-fat dairy products and eggs can help provide enough fat, but your baby may still fall short on iron and zinc. For these nutrients, you'll need to look to fortified cereals. According to the WHO, "Vegetarian diets cannot meet nutrient needs at this age unless nutrient supplements or fortified products are used."[35] I talk more about fortified cereals in the next section.

On the extreme end of vegetarianism are vegan and macrobiotic diets. Macrobiotic diets emphasize whole grains, legumes, nuts, seeds, seasonal fruits, and fermented foods, and avoid meat, eggs, and dairy products.

A study in the Netherlands found that babies and children raised on a macrobiotic diet had ubiquitous nutrient deficiencies, stunted growth, muscle and fat wasting, and slower psychomotor development.[36] A review of vegetarian diets for children noted that "the more restricted the diet and the younger the child, the greater the risk for deficiencies."[37] If you can, I recommend including meat, fish, eggs, and dairy in your baby's diet.

A ROLE FOR INFANT CEREALS AND OTHER GRAINS

Cereals are traditional first foods for babies around the world, and in the developed world, fortified infant cereals have been a nutritional cornerstone for babies for several generations. I've already discussed the role of meat and eggs in infant diets and the evidence supporting these foods as good nutrient-dense choices for first foods for most babies. If you're incorporating these foods into your baby's diet daily, she probably doesn't need fortified cereals. Likewise, fortified cereals aren't necessary for formula-fed babies, because formula is already fortified to meet babies' nutrient requirements through the first year. Still, fortified cereals can be useful for many breastfed babies. In this and the following sections, I discuss the advantages and disadvantages of fortified cereals, as well as grains in general, address some of the modern myths about grains, and give you the information you need to determine the role of grains in your baby's diet.

Let's start with what is in the typical infant cereal. At my grocery store, the plain old Gerber rice cereal is mostly rice flour—essentially a finely ground white rice. It's a refined grain, meaning that the bran and germ of the grain have been milled away, along with most of the nutrients, leaving just the starchy endosperm. (Brown rice is an example of a whole grain, since it still has the bran and germ intact. I discuss whole grains later.) Added to the rice flour is a bunch of supplementary vitamins and minerals (including a lot of iron), in part to replace those that were removed in refining, but also to make rice cereal a more nutrient-dense food to meet the specific needs of babies. In effect, rice cereal is a nutritional supplement carried by otherwise nutrient-poor refined flour, though reconstituting it with breast milk or formula adds protein and fat to make the meal more complete.

There's no doubt about it: baby cereal is a processed food. Why not serve our babies whole foods, foods our great-grandmothers would recognize, with more interesting tastes and textures? That is, after all, how we hope our babies will eat as they grow into children and adults.

This argument may have some merit, but let's consider the advantages of fortified cereals. Most obviously, fortified cereals are a good source of iron. Just two servings of baby cereal a day can nearly meet your baby's iron requirement, and studies show that fortified cereal prevents iron deficiency in breastfed babies. For example, one study found that in babies receiving a fortified cereal, just 2.5% developed iron deficiency. In contrast, in a control group of babies fed solids completely at the parents' discretion, 14% became iron-deficient.[38]

Meat may be the better way to get iron to your baby, but because so few babies eat meat regularly, fortified cereals play a vital role in preventing iron deficiency and anemia among this vulnerable age group. And compared with meat, cereals are often a more practical choice. They're inexpensive, convenient, and safe. They are stable to store at home or on the go, and you can make up just a tablespoon or two at a time, which is perfect for babies just starting to eat small amounts of solid foods. It is easy to adjust the texture, allowing you to start with very soupy cereal or to make it so thick that the baby can feed herself. I even incorporated fortified cereals into things like pancakes and muffins to increase the iron content in some of Cee's favorite finger foods. Cereals can be made up with breast milk or formula, providing a familiar taste bridge from milk to solids. One study found that breastfed babies were more likely to enjoy infant cereal when it was made up with breast milk, and the same is probably true for formula if it is the familiar milk for the baby.[39]

There has been some recent concern about arsenic in rice products, including infant cereals. Although I think rice cereal can be safely included in your baby's diet occasionally, it's a good idea to feed different types of grains, including oats, barley, and wheat (all of which come in fortified versions). In addition to reducing arsenic exposure, this gives your baby more variety in taste and texture. (See appendix H for more information about arsenic in rice.)

Whether it is your baby's major source of iron or an occasional supplement, fortified cereals may have a place in your baby's diet. If you do feed your baby commercial infant cereal, look for a brand that fortifies with both

✲ GOOD SOURCES OF VITAMIN C FOR BABIES

FOOD	*VITAMIN C (MG/SERVING)*
Fruits (raw)	
Kiwifruit	42
Citrus fruits (oranges, clementines)	25-35
Strawberries	22
Pineapple	20
Mango	15
Cantaloupe	14
Vegetables	
Bell pepper (raw, any color)	30-68
Broccoli (cooked)	25
Cauliflower (cooked)	14
Kale (cooked)	13
Sweet potato (baked, mashed)	10

Source: U.S. Department of Agriculture, Agricultural Research Service, "USDA National Nutrient Database for Standard Reference, Release 26," Nutrient Data Laboratory Home Page, 2013, http://www.ars.usda.gov/ba/bhnrc/ndl.

Note: Vitamin C amounts are given for a one-quarter cup serving of chopped fruit or vegetable (about 40 g). Aim to include at least 25 mg of vitamin C with a meal containing nonheme iron.

iron and zinc. (Almost all are fortified with iron, but not all companies have adopted zinc fortification.) Serve infant cereal with a good source of vitamin C, because it enhances the absorption of nonheme iron severalfold.[40] Some cereals are fortified with vitamin C; otherwise, pair the cereal with a vitamin C–rich fruit or vegetable. In addition to infant-specific cereals, many other hot and cold cereals found in your grocery store are fortified with iron. However, the iron source used in these cereals is less bioavailable than that used in infant cereal.[41] If cereal is your baby's major source of iron, infant cereals are a better choice.

WHAT ABOUT WHOLE GRAINS?

Whole grains are generally considered to be nutritionally superior to their refined counterparts. The whole grain includes not only the starchy endosperm but also the bran and the germ, which are naturally rich in vitamins, minerals, phytochemicals (a general name for chemical compounds produced by plants, some of which may be beneficial), and fiber. They also include some protein and healthy fats, giving us a nicely balanced package of nutrients. Whole grains are even reasonably high in iron. However, whole grains, like legumes and nuts, are also high in chemical compounds called phytates, and these bind iron during digestion, inhibiting absorption. Phytates are concentrated in the bran of cereal grains, and the higher the level of phytates, the lower the iron bioavailability.[42] Thus, from an infant nutrition standpoint, whole grains may be a sort of trade-off. They're better nutritionally in lots of ways, but refined cereals are usually a more bioavailable source of iron for babies.

Many baby food companies are nonetheless taking advantage of the popularity of whole grains and marketing whole-grain fortified infant cereals, although it isn't clear what effect phytate levels may have on iron bioavailability in these products. A recent survey of phytates in infant foods deemed the whole-grain cereals marketed in the United Kingdom and Denmark to have phytate levels "unsuitable for infants and young children."[43] However, a small U.S. study compared the iron status of babies fed a regular refined cereal and a whole-grain version, both fortified with iron.[44] Although the whole-grain cereal had nearly five times the phytate level of the refined version, there was no difference in babies' iron status in the two groups. That's encouraging, but this question needs to be studied in more depth.

Some parents avoid the commercial cereal products altogether and make their own whole-grain cereals. Of course, these aren't fortified with iron, but assuming that your baby has other good sources of iron, homemade whole grains are a nutritious and well-liked food for babies. You can take some steps at home to reduce the phytate levels in whole grains. Soaking grains and then rinsing with fresh water before cooking can decrease phytates; one study found that a six-hour soak in warm water decreased phytate levels by about half.[45] Germination of grains and legumes prior to cooking can similarly reduce phytates, and fermentation results in as much as 90% lower phytate levels.[46] Of course, all of these strategies take your time and energy, as well as a certain degree of dedication. And they remind me that while we might strive to avoid "processed foods," some processing—whether by soaking whole grains or refining them—can also have benefits when it comes to availability of nutrients.

BUT I HEARD THAT BABIES CAN'T DIGEST STARCH!

Adding to cereal's popularity slump, there's a funny rumor circulating around "crunchy" parenting websites that babies can't digest starch. Hence, some parents are choosing to avoid grains and other starchy foods (think root vegetables, winter squash, beans, and some fruits) until their children are 1 or 2 years of age. There is a truthful origin to this claim, but it has been grossly misconstrued.

Starch is made up of long, branching chains of glucose. When we eat starch, we need to break up those chains to liberate the glucose, making it available for absorption into our bloodstream so that it can be used as an energy source for cells. An enzyme called amylase, secreted both in saliva and from the pancreas into the small intestine, is responsible for most of our starch digestion. It turns out that newborns have little to no pancreatic amylase activity, and although this activity increases slowly over the first months of life, it doesn't reach adult levels until well into childhood.[47]

This deficiency in pancreatic amylase among babies puzzled the scientists researching it in the 1960s and 1970s. For one thing, babies appeared to digest starch just fine. When researchers fed starch to infants just a few months old, very little of it came out in their diapers.[48] And the infants didn't have symptoms of carbohydrate malabsorption, like diarrhea, nausea, cramping, bloating, and gas. (Think of what happens if someone who is lactose intolerant—lacking the enzyme to break down lactose—drinks milk.) These symptoms weren't apparent in young babies eating starchy infant cereals, which in the United States in the 1970s were usually introduced by 1 to 2 months of age.[49] And this approach to infant feeding wasn't unique to the United States. Ethnographic reports are filled with examples of starchy first foods for young infants around the world: millet flour at 3 months in Tanzania, corn porridge at 3 months in Zimbabwe, beans and rice at 4 months in Brazil, a little butter and flour at 3 *days* in Bhutan, rice mash at 3 weeks in Nepal, and pre-chewed taro root at 2 weeks in the Solomon Islands.[50] If all these little babies were eating starch, how were they digesting it without pancreatic amylase?

Babies seem to have a few ways around this. For one, babies make a lot of saliva (otherwise known as drool), and their saliva contains lots of salivary amylase. By 6 months of age, concentrations of salivary amylase in babies are near adult levels.[51] This amylase starts to break down food starch as the baby gums food around in her mouth, and it also seems to survive the trip through the acidic environment of the stomach and to continue its work in the more neutral small intestine.[52]

There is also a lot of amylase in human breast milk, 25 times more than that found in raw cow's milk.[53] It is highest in colostrum and decreases slowly during infancy, as the baby's salivary and pancreatic amylases increase.[54] Like salivary amylase, breast milk amylase appears to survive the acidic stomach, in part because it is protected by other proteins in breast milk.[55] This is one good reason to use breast milk to mix baby cereal, and amylase has been shown to be stable in breast milk even after freezing and thawing.[56] Finally, another enzyme, called glucoamylase, seems to help with starch digestion in babies. Glucoamylase is made by cells lining the walls of the small intestine, and it is very active in infants.[57]

Even after all of that amylase activity, some starch does escape the small intestine undigested, passing on to the colon.[58] This happens to some extent in adults too, and it isn't necessarily a bad thing. The beneficial oligosaccharides (short chains of sugars) found in breast milk, as well as dietary fiber from plant foods, also pass through the intestine undigested. Bacteria in the colon ferment these undigested carbohydrates, all as part of the symbiotic relationship between us and our gut microbes. The microbes benefit from this food supply, and it is becoming more and more apparent that the microbes benefit us as well. The end products of microbial fermentation in the colon are short-chain fatty acids, which can improve nutrient absorption, enhance gut health, and even be used as a source of energy for both the microbes and their human host.[59] Babies and toddlers may have faster colonic fermentation of starch than adults, perhaps representing an important pathway for them to fully capture the nutrients in their food.[60] The addition of complex carbohydrates, including starch and fiber, to the diet of older babies and toddlers might help to develop these beneficial microbes.[61]

When I hear about parents choosing to avoid grains and other starchy foods for the first year or two of their babies' lives, I feel concerned, for a couple of reasons. For one thing, as I noted in the previous chapter, there seems to be a critical window at around 6 months of age for introducing grains, including wheat. Delaying their introduction much beyond that time is associated with increased risk of wheat allergy, celiac disease, and type 1 diabetes. For another, I am concerned about our culture's ongoing obsession with restrictive diets. Barring allergies and intolerances, eating a variety of foods from all the food groups pretty much ensures that you'll meet your nutrient requirements without even trying, and I think this is a valuable practice to teach to our kids. Eating a diverse diet allows us all to relax and enjoy our food with the people we love, babies included.

LOVING YOUR FRUITS AND VEGETABLES

The most important solid foods for your baby are iron- and zinc-rich foods like meat or fortified cereal, but fruits and vegetables also play important roles. The first is related to iron. Vitamin C increases the absorption of nonheme iron from foods as much as twofold to sixfold, so including a fruit or vegetable high in vitamin C helps your baby get more iron out of foods like cereals and beans.[62] Fruits and vegetables are also good sources of trace minerals, folic acid, and B vitamins, as well as fiber, which is helpful for development of healthy gut bacteria and regular poops.

Beyond nutrition, there is a benefit to exploring lots of fruits and vegetables during late infancy. For one thing, babies this age are open to trying new foods, and they readily accept most flavors.[63] Randomized controlled trials have shown that when babies are offered a variety of fruits and vegetables daily, they become more likely to accept a new food.[64] Plus, greater dietary variety is associated with more balanced nutrition, and as a parent, you hope to establish this pattern in your child from the start. If your baby is hesitant about accepting new foods, remember that babies sometimes need eight or more tastes of a new food, in different meals, before they learn to love it.[65] Without putting pressure on your baby, continue to offer tastes and model enjoyment of these same foods yourself.

⁎ SHOULD YOU CHOOSE ORGANIC OR CONVENTIONAL FOODS?

Organic farming is usually better for the environment and for farm workers and results in lower levels of synthetic pesticide residues on food. But organic produce costs more, is no more nutritious than conventional produce, and can also be contaminated with organic pesticides (although not synthetic, these are not necessarily less dangerous). At this time, there is no convincing evidence that eating conventional produce is a threat to children's health. The U.S. Department of Agriculture's annual pesticide monitoring program for conventional produce consistently finds that pesticide levels are hundreds to millions of times lower than the daily limits set by the U.S. Environmental Protection Agency. Commercial baby food, whether organic or conventional, carries very little to no pesticides. Buying organic is a good way to support organic farming systems, but it is unlikely to affect your child's health.

There is one thing we know for sure: a diet rich in fruits and vegetables, however they are grown, has many health benefits. The best nutrition is found in the freshest produce, grown in your own garden or purchased from local farms. Buying local also gives you the benefit of talking directly with farmers about their spraying practices, supporting your local economy, and showing your kids where their food comes from.

C. K. Winter and J. M. Katz, "Dietary Exposure to Pesticide Residues from Commodities Alleged to Contain the Highest Contamination Levels," *Journal of Toxicology* 2011 (2011): 1-7.

For most babies, open-mindedness to trying new foods won't last. At some point in their second year of life, most become more skeptical of new flavors and accept fewer new foods. This may be true of lots of foods, but the inherent bitterness of vegetables makes them a common victim of picky eating. Trying lots of fruits and vegetables during the "honeymoon" phase of eating in late infancy may not prevent their rejection later, but it might help. It's also just plain fun to eat with a baby who joyfully eats everything, from broccoli to Brussels sprouts.

A generation ago, parents were given precise advice about what foods to introduce when, and in what order, but there is really no evidence to back a schedule like this. For example, introducing fruits before vegetables won't get in the way of your baby liking vegetables.[66] Cooking fruits and vegetables can improve their digestibility and sometimes decrease the chance of triggering an allergic reaction, so it may be wise to start with cooked versions (which are softer in texture anyway) and then progress to the raw form. Whether you choose organic or conventionally grown foods for your baby is a personal choice, but there is little evidence that this will affect your child's health.

SUGAR, SPICE, AND A PINCH OF SALT

A 2008 survey of U.S. babies found that 17% of 6- to 9-month-olds and 43% of 9- to 11-month-olds consumed desserts, sweets, or sweetened beverages at least once a day.[67] This is concerning. Sugary foods displace much-needed nutrients during this period of rapid growth and development, and they promote the development of cavities. Babies like sweet foods, and feeding them sweets only increases their preference for them.[68] A preference for sweet taste may have been evolutionarily beneficial in an environment where sweetness might signal a good, safe source of calories.[69] But these days, our babies are usually raised in environments where calories are plentiful, and we have to lead by example to introduce them to the complex tastes of a variety of foods. Don't hesitate to feed fruits, which are naturally sweet but also come in a package that includes beneficial fiber, vitamins, and minerals. But for babies under 1 year, avoid feeding foods with added sugar, except in small amounts as an occasional treat. There's also no reason to give babies juice; whole fruits provide more nutrients. If you do give your baby juice, dilute it with water and give it in a cup, not a bottle. Bottles make it too easy for your baby to drink a lot of juice, and this is associated with greater risk of cavities. Finally, avoid honey completely during the first year of life. The toxin that causes botulism is often found in honey, even when baked into foods, and honey consumption has been associated with infantile botulism.[70]

Likewise, there's no reason to add salt to your baby's food. A Dutch

⁎ THE INS AND OUTS OF COMMERCIAL BABY FOODS

It's easy to make your own baby food. All you need is a blender or food processor, or even just a fork for mashing. But what if you're short on the time or the desire to make your own? Are store-bought foods a good option?

It depends on the product. If you choose simple purees that contain just real fruits or vegetables, then store-bought foods are fine, though more expensive than what you could make in your own kitchen. However, be wary of ingredients like "puree concentrate" or "fruit juice concentrate." Although these ingredients are made from real fruits, they are processed to remove water and often fiber, which ends up concentrating the sugars. Check labels carefully.

Baby food companies like you to think that you need to introduce

study put some babies on a low-salt diet for the first six months of life, and at 15 years of age, these children had lower blood pressure than those in a control group, despite having similar salt intake at that age.[71] I started adding small amounts of salt to Cee's vegetables soon after her first birthday, around the time she started showing a little bit of pickiness and rejecting some vegetables. In my opinion, a little bit of salt does wonders for vegetables, but babies don't need it and will readily accept most veggies without it. Having the ability to control sugar and salt content is one reason to make your own baby food, but with careful label reading, you can find good options among commercial baby foods.

The focus of feeding your baby is to provide exposure to lots of different tastes, textures, and flavors. As your baby begins to eat solid foods, you'll want to start with single-ingredient foods so that if she has an allergic reaction, you'll know the culprit. However, once your baby has tried a food a few times, this is a great time to introduce spices to her palate. As far as I can tell, there aren't any studies on feeding spicy foods to babies, but given the variety of flavors fed to babies around the world, it's hard to imagine that offering spicy food is a bad thing. Just start with small amounts and be observant; your baby will let you know if the flavor is too strong for her.

foods to your baby in predetermined "stages," which linger in various pureed variations for way too long. Most babies need pureed foods only for a few months, if at all (some skip them entirely), and you should follow your baby's lead to advance to more complex textures and self-feeding.

The real rip-off in baby foods is the meals and snacks marketed for older babies and toddlers. These often have added sugar, salt, and filler ingredients. For example, I found a turkey stew meal that includes an ingredient called "cooked white turkey meat chopped and formed," which, as is clarified in parentheses, contains not only turkey meat and broth but also modified cornstarch, tapioca starch, and salt. These meals are expensive and nutrient-poor, and they don't help your baby learn to eat real food.

THE MOST IMPORTANT PART OF FEEDING: IT ISN'T THE FOOD

In this chapter, I've summarized the foods that are most nutritionally beneficial to growing babies. In an ideal world, I think babies do best eating meat, rich egg yolks, a mix of different types of grains, and a variety of fruits and vegetables. However, I live in the real world, just like you, and I don't even follow my own advice all the time. For example, even though meat may be a better source of iron, I still don't hesitate to tell busy parents that fortified cereals are a great way to ensure that their baby gets enough iron—because it's true; these cereals work just fine. And some parents get more joy out of preparing foods for their babies than others. There's no shame in shortcuts.

We humans are flexible omnivores; over the ages and around the world, we have thrived on very different diets. The nutritional principles in this chapter can help you figure out the right foods for your baby, but in practice that might look very different from household to household, depending on dietary preferences and traditions. The goal is to bring your baby to the family table and explore a variety of nutrient-dense foods that fit your life.

I also want to leave you with a note of caution: beware of making nutritional perfectionism your goal. Even if you're committed to providing the best foods for your baby every day, your baby may have her own ideas about what that means. You might be determined that your baby should eat salmon, for example, and she might turn up her nose at it. You have to honor that. Keep offering exposure and modeling happy eating of good foods, but don't push your baby to eat something she doesn't want. If you're worried about specific nutrients, find alternative sources and talk to your pediatrician. And then relax, and enjoy feeding and eating with your baby.

EPILOGUE

The room was hushed. I watched on the ultrasound monitor as strokes of gray gave way to a dark space—a gestational sac. It looked empty at first, but within seconds, an oblong smudge came into view at the edge of the darkness. And then, within that smudge, came a little flutter, barely perceptible but persistent. "There's the heartbeat," the ultrasound tech said. "It probably just started beating today, or yesterday." I was pregnant, but just barely.

This news came as I was writing the last chapter of this book. I pushed through waves of nausea, fatigue, and anxiety to finish writing and editing the manuscript. All the while, as I struggled with the many words in the book, I chanted just two, over and over, in my head: "Be there. Be there. Be there." Sometimes I'd place a hand on my flat belly, look down from my computer, and repeat, "Be there. Please." It wasn't very scientific, but it was all I had.

As I write this, on the day that I'll send this book manuscript to the publisher, I'm about 14 weeks pregnant. It's still there.

The journey to this pregnancy has been just as long as the journey of this book. The two have been intertwined, and both have taken me to the intersection of parenting and science.

On the day I signed the contract committing to write the book, almost two years ago, my period was one day late. A couple of days later, a home pregnancy test confirmed that I was pregnant. Cee would have a little brother or sister at just about the sibling interval we had hoped for. And I would tackle this book with the looming deadline of a due date and lots of personal motivation to answer all of my own questions before the arrival of another baby.

But that wasn't how it turned out. That pregnancy ended in miscarriage, a profound and intangible loss. What I mourned the most was the vision of the family I wanted: the baby born in summertime, the doting big sister that Cee would be, and the personal growth for my husband and me as we adapted from parenting one to two. With the loss of that pregnancy and two more miscarriages that followed, there also came a feeling of loss of control. I couldn't control every process in my body, the timing of our next baby, or the age gap between our children. Things were not going to turn out as I had planned. It was humbling and disheartening.

This is really a magnification of what parenting is every single day. I don't know anyone who finds that having a baby is exactly what they envisioned. Some parts may turn out as we had hoped, but others will be much harder, and a few will be surprisingly easy. We can't pick and choose the challenges, and we can't always predict how we will react to the stress and uncertainty of parenting. And none of us will experience it in the same way. This fact was front and center in my mind as I wrote this book, and it gave me more empathy for all the challenges and successes we each have with parenting.

Science continued to be a refuge for me as I worked on the book, even as I recognized its limitations. I could set aside my personal struggles and lose myself in a fascinating world of babies and parents, seen through the lens of science. It gave me a greater appreciation for the complexities of this task of parenting and for the many ways that good parenting takes shape across different families.

Now I get the absolutely best reward I could hope for. If all goes well, I'll have a new baby to care for by the time this book is published. I'll get to watch that baby grow and develop and fit into our family. Maybe I'll even get to use a few of the things that I learned in the process of writing the book. What is certain is that this baby, like all new babies, will bring new challenges—and *lots* more questions.

APPENDIXES

APPENDIX A

Are the Ingredients in the Newborn Vitamin K Shot Safe?

In chapter 3, I discuss the evidence for giving the vitamin K shot at birth. Despite its proven benefits, some parents opt out of giving their newborns this shot. When asked why, one of the common concerns is that the shot contains "toxins."[1] I've carefully evaluated each ingredient and concluded that this fear is unfounded. The vitamin K shot is safe. All of the ingredients serve important roles in the efficacy and safety of the shot, and none pose toxicological concerns in the amounts used.

Anytime you want to know whether a chemical is safe or effective, it is critical to look at the amount used, or the dose. A life-saving drug, for example, could be worthless if you didn't take enough of it. It could also be toxic if you took too much. For the vitamin K shot, there *are* toxicological concerns with several of the ingredients, and these concerns are listed on the package insert for the vitamin K shot. However, in all cases, these were based on incidents using large, often continuous, *intravenous* doses of these ingredients, quite different from the one-time intramuscular dose of the vitamin K shot.[2]

Two types of vitamin K shots are used in the United States. Both contain phytonadione, another name for vitamin K_1, but they contain different "inactive" ingredients. Most hospitals use the preservative-free version for newborns, so let's start with the ingredients in this one.

* *Propylene glycol* acts as a solvent to keep the vitamin K in solution. (Vitamin K is fat-soluble, so just as oil and water don't mix, it can't simply be dissolved in water or saline for the injection.) Propylene glycol is commonly used in foods, supplements, and other medications, and in the amounts used, it is safe. Don't confuse propylene glycol with ethylene glycol, a much more toxic chemical used in antifreeze. (If antifreeze is labeled "nontoxic antifreeze," it's probably made with propylene glycol.) According to the World Health Organization, propylene glycol is safe to consume in amounts up to 25 mg/kg of body weight per day. The vitamin K shot includes 10.4 mg of propylene glycol, or about 3.3 mg/kg for a 7-pound newborn.[3]
* *Polysorbate 80* is made from a plant-based sugar alcohol (sorbitol) and a fatty acid (oleic acid). It's often used in ice cream to make it smoother,[4] and it serves a similar purpose in the vitamin K shot. It is an emulsifier, helping fat-soluble vitamin K stay in solution. Polysorbate 80 was probably the culprit when an intravenous vitamin E product proved to be toxic in premature babies (an incident included on the package insert for the vitamin K shot), but these babies received more than 70 mg/kg *daily* for weeks and even months.[5] The one-time vitamin K shot contains 10 mg of polysorbate 80 (about 3 mg/kg for a 7-pound baby), and there is no reason to believe that this amount is harmful to newborns.[6]
* *Sodium acetate anhydrous* (0.17 mg) helps to maintain a neutral pH in the vitamin K shot. There are no toxicity concerns with this small amount.[7] This ingredient does contain a small amount of aluminum, which is noted on the package insert, but the amount is no more than 0.05 µg (a µg is 1/1000th of a milligram).[8] At birth, a newborn's body already contains 8,000 times this amount of aluminum.[9]
* *Glacial acetic acid* (0.002 ml) is a weak acid and the main component of vinegar. It is used to adjust the pH and has no safety concerns as used in the vitamin K shot.[10]

The second type of vitamin K shot used in the United States is less commonly used for newborns, but it is also safe. It contains the following ingredients:

* *Benzyl alcohol* prevents bacterial growth in the vitamin K shot. The

WHO sets the safe daily intake of benzyl alcohol at no more than 5 mg/kg. The vitamin K shot is well below this level; it contains 4.5 mg, or about 1.4 mg/kg for a 7-pound newborn. The vitamin K package insert warns that benzyl alcohol can cause toxicity in newborns, but this requires much larger, daily, intravenous amounts (more than 99 mg/kg/day).[11]

* *Polyethoxylated castor oil* acts as an emulsifier and solvent to keep vitamin K in solution and is used in many other drugs for this purpose. Very rarely, this ingredient has caused an anaphylactic reaction in patients receiving large doses by intravenous infusion, but there are no reported cases of this effect in newborns receiving the intramuscular vitamin K shot with this ingredient.[12]

* *Dextrose monohydrate* is a fancy name for glucose, the simple sugar that circulates in our blood and fuels our cells. Much larger doses are routinely given to newborns who have low blood sugar. There are no safety concerns with this ingredient.[13]

A third product, called Konakion MM, made by Roche Pharmaceuticals, is licensed for both oral and injectable use in Europe. First introduced in 1994 in Switzerland, this product contains glycocholic acid (a bile salt) and lecithin (a phospholipid) to disperse the fat-soluble vitamin K into small droplets, similar to the mixed micelles (hence the MM in the name) that are part of the normal digestion of fats and fat-soluble vitamins. The idea behind this preparation was that it would be more absorbable via the oral route, even for babies with liver disease and limited bile salts, but it hasn't turned out to be much more effective than the other preparations at preventing late vitamin K deficiency bleeding (VKDB) in these babies.[14]

The package insert for the vitamin K shot notes that it can cause jaundice (hyperbilirubinemia), and this is often mentioned in anti–vitamin K articles. However, these cases occurred in babies receiving 10 to 20 mg doses of a synthetic, water-soluble form of vitamin K, which is no longer used. Jaundice has not been observed in babies who received the recommended dose of 1 mg or less.[15]

APPENDIX B

Why Is the Hepatitis B Vaccine Recommended at Birth?

In the United States and many other countries, it is recommended that babies get their first dose of the hepatitis B vaccine at birth (depending on risk factors, either within 12 hours of birth or before discharge from the hospital). Another dose is given at around 2 months and a third dose between 6 and 18 months of age.[1]

The hepatitis B virus (HBV) infects the liver. An acute HBV infection causes nausea, vomiting, jaundice, loss of appetite, and abdominal pain, but most people recover without lasting effects. Much more serious is a chronic HBV infection. Although a person may be asymptomatic for years (and thus not know that he or she is carrying the disease, so inadvertently infecting others), chronic HBV infections may eventually cause cirrhosis or liver cancer and are responsible for about 3,000 deaths each year in the United States. Unfortunately, babies and young children are at very high risk of developing chronic infections because they are unable to clear the virus. Ninety percent of infected infants develop chronic infections, and of these, 25% will die prematurely due to complications of the disease.[2]

HBV is transmitted through blood, semen, and saliva. Some people mistakenly assume that sexual contact is the only way that HBV can be transmitted from one person to the next, but this is not the case. Most HBV-positive moms pass the virus on to their newborn babies, although vaccination at birth will prevent the majority of these infections. For this reason, women should be tested for HBV during the first trimester of pregnancy so that their babies can be vaccinated and treated for HBV soon after birth.[3] However, the women at greatest risk of being HBV-positive are the least likely to have prenatal care, and infection can also occur between the first-trimester test and the time of giving birth.[4]

There are other ways for babies to be exposed to HBV. The virus can live for more than seven days at room temperature; it can easily be passed from one person to the next on a washcloth or toothbrush. It has been passed from one child to another through biting in a day care setting.[5] I thought about HBV when, as a toddler, Cee started "discovering" litter such as candy wrappers and empty water bottles at the park. The birth dose of HBV vaccine protects babies from the start, should they be exposed to the virus.

The HBV vaccine is highly effective, inducing immunity to HBV in 98% of infants who complete the series.[6] Because it prevents chronic liver infections, which cause liver cancer, the HBV vaccine is an anticancer vaccine. Before the vaccine, about 300,000 people, as many as 24,000 of them children, were infected with HBV each year in the United States.[7] With the universal vaccination of infants, recommended in 1991, the incidence of acute hepatitis B in children less than 12 years old has decreased by 94%, and the incidence in the overall population has decreased by 75%.[8] Chronic HBV infections are more difficult to track because they are often asymptomatic. However, since they can take decades to cause illness, the greatest benefits of HBV vaccination—reduction in chronic infections—may not yet have been detected.[9]

The HBV vaccine is very safe. It can cause minor soreness at the site of injection and a low-grade fever. Very rarely, it can cause a severe allergic reaction (1 in 600,000 doses), but if this happens, a health care provider can rapidly respond.[10]

Based on the evidence for the safety of the HBV vaccine and the risk of infection in early infancy, it makes sense for babies to be vaccinated at birth. Some families with HBV-negative moms and low risk factors in their immediate family might choose to wait a week to two months to give the first dose of the HBV vaccine. Although this still increases a baby's chances of becoming infected with HBV in the first few weeks of life, this is a reasonable choice, but one you should discuss with your pediatrician.[11]

APPENDIX C

Do We Give Too Many Vaccines Too Soon?

When my parents were born, in 1948, the only vaccines available and recommended were for smallpox and a combined shot for diphtheria, tetanus, and pertussis. That was it—protection from four diseases. By the time I was born, in 1980, the measles, mumps, and rubella (MMR) and the polio vaccines had been added to the schedule. Smallpox had been successfully eradicated, and the vaccine was no longer needed. I was immunized for seven diseases. In 2014, as I write this, the recommended schedule provides protection against 14 diseases, requiring as many as 27 doses in the first two years of life.[1] Your baby will often get several shots at a time, although combination shots have helped to reduce the total number needed. The sheer number of needles you see at a given well-child visit might make you wonder: can your little baby handle so many vaccines, in such quick succession, at such a young age?

As parents, we are most aware of the number of needles we see, but a more important measure of the vaccine "load" is the number of unique antigens—proteins or polysaccharides that stimulate an immune response. Over the past few decades, as new vaccines have been developed, they have been refined to contain fewer antigens at smaller doses, just enough to give an immune response. When I was a child, immunization against seven diseases required exposure to more than 3,000 antigens. Today's recommended schedule for babies protects against 14 diseases, and it does so with only about

ANTIGEN LOAD IN VACCINES RECOMMENDED FOR CHILDREN UNDER 2 YEARS OF AGE IN THE UNITED STATES, 1900-2014.

	1900	1960	1980	2014
Disease protection provided by recommended schedule	Smallpox	Smallpox Diphtheria Tetanus Pertussis Polio	Diphtheria Tetanus Pertussis Polio Measles Mumps Rubella	Diphtheria Tetanus Pertussis Polio Measles Mumps Rubella Hib* Varicella Pneumococcus Hepatitis A Hepatitis B Rotavirus Influenza
Total number of vaccines	1	5	7	14
TOTAL NUMBER OF ANTIGENS	~200	~3,217	~3,041	154-163†

Sources: F. DeStefano, C. S. Price, and E. S. Weintraub, "Increasing Exposure to Antibody-Stimulating Proteins and Polysaccharides in Vaccines Is Not Associated with Risk of Autism," *Journal of Pediatrics* 163 (2013): 561-67; P. A. Offit et al., "Addressing Parents' Concerns: Do Multiple Vaccines Overwhelm or Weaken the Infant's Immune System?" *Pediatrics* 109 (2002): 124-29; P. A. Offit, email to the author, March 13, 2014.

Note: The two vaccines with the largest number of antigens were smallpox and the whole-cell pertussis vaccine. Since global eradication of smallpox, the vaccine is no longer used. The whole-cell pertussis vaccine contained about 3,000 antigens, but in 1999 it was replaced with the acellular vaccine, containing only parts of the pertussis pathogen and just 2 to 5 antigens.

*Haemophilus influenzae type B.

†Several types of vaccines are licensed for some vaccinations, so the number of antigens depends on which vaccine is received.

160 antigens. Advances in science have given us the benefit of greater protection from infectious diseases *and* smaller, safer vaccines.

When a baby is given a vaccine, her immune system responds to those antigens, but this is only one of many tasks that it is handling on that day. From the moment of birth, a newborn baby is bombarded with bacteria and viruses, and she has to generate a range of effective immune responses to sort the good from the bad and to survive in a germy world. We know that one or five vaccines don't overwhelm a baby's immune system. For example, when a child that is already battling an ear infection, upper respiratory tract infection, or diarrhea is vaccinated, she is able to mount just as robust an immune response to the vaccine as a perfectly healthy child, and with no greater risk of an adverse reaction. Her immune system can handle it. And when combination shots are given, the immune response to each vaccine is no different from when the vaccines are given separately.[2]

If the recommended immunization schedule overwhelmed infant's immune systems, we would expect vaccinated babies to be less capable of responding to other infectious diseases. A randomized controlled trial in Germany tested this hypothesis, dividing 496 newborn babies into two groups, one receiving five vaccines at 2 months, and the other delaying those vaccines until 3 months.[3] Between 2 and 3 months of age, the unvaccinated babies were *more* likely to have symptoms of illness such as vomiting, coughing, runny nose, and rash. Clearly, the vaccinated babies' immune systems weren't overwhelmed; in fact, they appeared to have been strengthened in some nonspecific way that protected them from everyday sorts of pathogens.

A 2013 study specifically looked at the concern that the number of vaccines in the modern immunization schedule may be associated with a higher risk for autism. It tallied the total number of antigens in the vaccines given to children with and without autism, and it found no relationship between the vaccine "load" on the immune system and development of autism.[4] Another study found that children who received all of their vaccines on time in the first year of life scored the same or better on neuropsychological outcomes compared with children who had delayed or fewer immunizations.[5] Finally, when the Institute of Medicine evaluated the safety of the entire childhood immunization schedule, it found no evidence that the recommended schedule is linked to autoimmune diseases, asthma, hypersensitivity, seizures, learning or developmental disorders, or attention deficit or disruptive disorders.[6]

The concern that we give too many vaccines too soon is unfounded, but it has led some parents to vaccinate on a delayed schedule, spreading out shots to reduce the number given at one time or waiting until their baby is older for some shots. There are a few problems with this. First, there's no evidence that delaying or spacing out vaccines offers any benefit or is any safer. And, on the flip side, delaying vaccines has known risks. The recommended schedule is based on expert assessment of the research on vaccine safety and efficacy, and it is specifically designed to vaccinate children when they are able to respond appropriately to the vaccine and when they are most vulnerable to the disease. Delaying vaccines increases the amount of time that your child is susceptible to getting sick or passing a disease on to a newborn baby or person who has a medical reason for not being able to get vaccinated.

As children grow and develop, so do their immune systems. It seems intuitive, then, that a child might be able to respond better to a vaccine if it is given at an older age. However, this is a simplistic way of looking at things. For example, febrile seizures are a known and rare side effect of the MMR vaccine, occurring in 1 in 3,000 vaccinated children. Febrile seizures aren't dangerous and don't cause long-term damage to the child, but understandably, they are scary to parents. A 2013 study of more than 840,000 children found that compared with those vaccinated at the recommended time (12 to 15 months), children that got the MMR shot at 16 to 23 months had double the risk of a seizure.[7] In other words, delaying the MMR shot actually increased the risk of side effects. This could be because the older age group did indeed have a more rigorous immune response to the vaccine, but if this increases side effects as well, then later vaccination isn't necessarily the better option. Giving the MMR shot at the recommended time ensures an appropriate immune response—enough to provide immunity to measles, mumps, and rubella—at an age for which the side effects are well studied. Deviating from the schedule puts your child in uncharted territory for both safety and efficacy of the vaccine.

Finally, spacing out your child's vaccines has some practical implications. It means more trips to your doctor's office for shots, and if you're trying to split up combination shots, it increases the total number of shots given. That could mean more overall stress for your baby.[8] Each visit to the doctor's office also increases your baby's risk of coming into contact with another sick child, who could make her ill. In other words, spacing out your baby's vaccines just adds up to more stress, more risk, and no known benefit.

APPENDIX D

Do Vaccines Cause Autism?

The purported link between vaccines and autism is probably the vaccine-related concern that has garnered the most attention in the media in recent years. Such a link has been thoroughly investigated by scientists around the world, and study after study has found no evidence that vaccines cause autism. Still, this concern is often cited by parents as a reason why they hesitate to vaccinate their children.[1] So, let's take a closer look.

There are two historical origins of theories about vaccines and autism. The first is a 1998 paper published in the *Lancet* by London gastroenterologist Andrew Wakefield.[2] The paper was a case series of 12 children, all of whom had symptoms of digestive problems and concerns about their neurodevelopment, many (but not all) of the children having been diagnosed with autism. (Recall from chapter 1 that case reports and case series provide the lowest quality of scientific evidence.) The paper didn't provide any evidence for a link between the MMR vaccine and children's medical symptoms, but it did allude to this possibility, and Wakefield voiced his concern about the MMR shot in press conferences following its publication. This one paper was successful in getting a disproportionate amount of media attention, given its weak findings, and fear about the MMR vaccine led parents to opt out of the MMR vaccination, particularly in the United Kingdom. Unfortunately, measles outbreaks have followed. In 1998, when Wakefield published his paper, there were 56 cases of measles in England and Wales. Fifteen years later, in 2013, there were 1,843 measles cases, many of them among older kids and teens who didn't receive the vaccine during the autism scare. Several children have died of measles in recent outbreaks in the United Kingdom and Ireland.[3]

A number of concerns arose about Wakefield's research, and an investigation found him guilty of ethical, medical, and scientific misconduct. Wakefield was stripped of his medical license, and the *Lancet* finally retracted his paper in 2010.[4]

A second autism-related matter focused on the use of the preservative thimerosal in vaccines. Thimerosal prevents bacterial contamination of multiple-dose vaccines, an important role in maintaining their safety. Thimerosal contains ethylmercury, and there was some concern that the total amount of thimerosal in the childhood vaccination schedule could cause neurological problems. In 1999, the U.S. Public Health Service and the American Academy of Pediatrics asked that thimerosal be removed from childhood vaccines (which meant switching to more expensive single-dose vials), not because there was any evidence that it was a problem, but because they thought they didn't have enough evidence to say for sure that it wasn't.[5] Subsequent research has shown that the ethylmercury in thimerosal is much less toxic than the methylmercury found in fish. Ethylmercury is metabolized and cleared by the body much faster, most of it being excreted in the stool, and the amount used in vaccines did not pose a risk to children's health. Regardless, thimerosal is no longer used in childhood vaccines in the United States, except for some types of flu vaccines.[6]

Both hypotheses—that autism was in some way related to the MMR vaccine or to thimerosal—have been thoroughly investigated and thoroughly rejected. These studies have been conducted around the world, from Canada to California, Finland to Japan, and they've looked at millions of children in multiple types of study designs. The Institute of Medicine, the Cochrane Collaboration, and others have carefully reviewed these studies and found no link between vaccines and autism.[7] A 2014 meta-analysis combined the results of all the high-quality studies of vaccines and autism published to date, including nearly 1.3 million children, and found no relationship between autism and MMR, thimerosol, mercury, or vaccines in general.[8]

We don't know exactly what causes autism, but we do know that genetics plays a big role, and prenatal factors may contribute.[9] Whatever the cause, most evidence to date indicates that autism starts in the womb. For example, a 2014 study published in the *New England Journal of Medicine* showed that children with autism have disorganization in the layers of cells in the cortex of the brain, changes that must have occurred during fetal development.[10] A 2013 paper in the journal *Nature* reported that differences in social interaction begin to appear in babies later diagnosed with autism as early as 2 months of age, long before children receive their first MMR shot (around their first birthday).[11]

The vast amount of research on vaccines and autism—all finding no relationship—should have put these fears to rest years ago, but some parents still insist that their children's autism was related to vaccines. This is understandable. Children receive a lot of vaccinations during the first two years of life, and parents begin to notice the signs of autism during this time (although biologically, the differences that make a child autistic have most likely been present in that child from before birth). It's human nature for parents to look for some explanation or something to blame to help them make sense of a change in their child's behavior, and statistically, some parents are bound to notice these changes soon after a vaccination. But what parents are noticing here is a correlation, not evidence of causation. We need science to evaluate causation. We need careful, systematic study of many variables in many people. When scientists have looked at the autism question in this way, it is clear that there is no relationship between autism and vaccines.

It's good to know that vaccines don't cause autism. But, in all, this is a very sad story. A few scary stories were blown up in the media and planted seeds of fear, which then spread like wildfire on the Internet. Immunization rates went down, leaving kids vulnerable to disease and leading to outbreaks. Lots of energy and money were focused on the vaccine hypothesis, which never had much evidence supporting it in the first place. Those resources would have been much better spent on research to understand autism or on helping families support their autistic children.

APPENDIX E

Do Vaccines Increase a Baby's Risk of SIDS?

In the past, there were reports that the DTP shot, which included the diphtheria, tetanus, and whole-cell pertussis vaccines, might increase the risk of SIDS. Research showed that this is not the case. (Plus, the whole-cell pertussis vaccine was replaced with an acellular version in most countries.) Just as with the vaccines and autism theory, the SIDS theory took hold because of an association in time between SIDS and vaccines. Most SIDS deaths occur between 2 and 6 months of age, the same time that infants are getting several rounds of immunizations. Statistically, there is a good chance that a baby who died of SIDS was recently immunized, and for parents looking for a way to explain the inexplicable, vaccines might look like a reasonable culprit. This is a classic example of confusing correlation and causation. A recent meta-analysis combined the results of nine studies and found that, in fact, immunizations *reduce* a baby's risk of SIDS by about half.[1] We know that illness increases a baby's risk of SIDS, so this could be an effect of a decreased incidence of infections. An alternative explanation is that parents delay vaccinating babies when they are ill, such that vaccinated babies are a healthier group from the start and thus less likely to succumb to SIDS. Regardless, you can rest easy knowing that your baby's shots won't increase SIDS risk.

APPENDIX F

Should We Worry About Aluminum in Vaccines?

Small amounts of aluminum salts are present in some vaccines. They are used as adjuvants, substances that help to stimulate the immune response to a vaccine, often allowing smaller quantities of antigen or fewer doses to be used. Without aluminum, some vaccines are less effective, meaning that more people fail to develop immunity in response to the vaccine. Aluminum may also reduce adverse reactions to vaccines by slowing the release of antigens from the muscle, where the vaccine is injected, into the bloodstream. Aluminum salts have been used as adjuvants in vaccines for more than 70 years, and like every ingredient in the vial, they play an important role.[1] However, some parents are worried about the cumulative effect of aluminum in vaccines. What does the science say?

Aluminum is all around us. It's the most abundant metal in the earth's crust, and it is found in our drinking water, our food, and the air we breathe, in products like sunscreen and antiperspirants, and in antacids and other medicines. At birth, a newborn already has some aluminum in his body (~0.4 mg), and this "body burden" builds slowly across the lifespan with exposure to aluminum from a variety of sources. By 6 months, an infant will have consumed 10 mg of aluminum if breastfed, 40 mg if fed regular formula, and 120 mg if fed soy-based formula. Starting solid foods means even more aluminum.[2]

It's true that too much aluminum can have harmful effects on health, but as with any chemical, we have to look at the *amount* of the dose and the *route* of exposure to assess its toxicity. Over the first year of life, an infant immunized according to the recommended schedule will receive, at most, 4.2 mg of aluminum from vaccines, with no more than 1.2 mg at a time. The aluminum salts used in vaccines are insoluble, so they dissolve and enter the bloodstream very slowly after injection. For a long time, blood tests were unable to show an increase in aluminum concentrations following vaccination. More recent experiments using sensitive mass spectrometry equipment show an increase of less than 1% in blood aluminum levels after an immunization.[3] After this gradual absorption into the blood, much of the injected aluminum is filtered by the kidneys and excreted in the urine. A small amount contributes to the body burden, mostly stored in bone, but this is well below the safety limits established for infants.[4]

Many parents have been misled about aluminum in vaccines by Dr. Robert Sears, whose popular book encourages parents to space out shots, using one of his recommended alternative schedules. Dr. Sears cites concerns about too much aluminum as one of his reasons for condoning an alternative schedule. However, he based his aluminum calculations on studies of intravenous aluminum in hemodialysis patients and severely premature infants, both of which have impaired kidney function.[5] Intravenous administration of aluminum acts very differently from an intramuscular injection, as used for vaccines, because the aluminum in an IV injection is already dissolved and hits the bloodstream all at once.[6] Dr. Sears's aluminum calculations fail to take this into account, not to mention that he's basing them on patients whose natural pathway for aluminum excretion is compromised. There is no indication that the aluminum in vaccines is dangerous, and no reason to spread out vaccinations to reduce the amount of aluminum given at a time. (Plus, there are good reasons *not* to space out vaccinations; see appendix C.)

APPENDIX G

How Do We Know That Babies Need So Much Iron?

I've discussed iron a lot in this book. Iron nutrition is one of the most compelling reasons to delay cord clamping (chapter 2), start solid foods by 6 months (chapter 9), and choose iron-rich complementary foods for your baby (chapter 10). If you're like me, you might wonder how we know that babies need so much iron. How do we determine the iron requirement for babies, and how certain are we of its accuracy? These are great questions, and since iron is so important to babies, I thought I'd dissect the calculations that go into the iron requirement.

The WHO suggests that babies aged 6 to 12 months need to absorb 0.93 mg of iron from their food each day to meet their body's needs.[1] How did the experts on the WHO committee arrive at this number? To start, they looked at how much a baby grows during this period. As the baby grows, her blood volume increases, and iron is required to make more hemoglobin. Tissues, including the developing brain and the muscles working overtime as your baby learns to crawl, also grow in size and need more iron. For normal growth and development, the average baby needs about 0.55 mg of iron per day. In addition, 0.17 mg of iron is lost in feces, urine, and dead skin every day and has to be replaced. Add these two numbers together, and that brings us to 0.72 mg of iron needed each day by the average baby.

But this number is just an *average*. Some babies will need less iron and some will need more, depending, in part, on their size and growth

✲ WORLD HEALTH ORGANIZATION'S RECOMMENDED IRON INTAKES FOR INFANTS 6 TO 12 MONTHS OF AGE BASED ON BIOAVAILABILITY IN FOOD

DIETARY BIOAVAILABILITY	*15%*	*12%*	*10%*	*5%*
Iron requirement (mg/day)	6.2	7.7	9.3	18.6

Source: Joint FAO/WHO Expert Consultation on Human Vitamin and Mineral Requirements, and WHO Department of Nutrition for Health and Development, *Vitamin and Mineral Requirements in Human Nutrition*, 2nd ed. (Geneva: World Health Organization, 2005).

Note: With higher dietary bioavailability (as from heme sources), less dietary iron is required. A vegetarian diet has about 5% bioavailability, whereas a diet with frequent servings of meat has a bioavailability of 15% or more.

rates. We want *all* babies to get enough, so a buffer is added to the iron requirement based on the variability in growth rates, and that brings us up to the WHO's recommended 0.93 mg of absorbed iron per day for babies aged 6 to 12 months, which is estimated to meet the needs of 95% of all babies.

This number, however, is *absorbed* iron—the iron that actually crosses from the digestive tract into the blood to be used by the body. Iron bioavailability is highly variable, depending on the food source and other components of the diet. The best iron bioavailability is found in heme sources such as meat, poultry, and fish; about 25% to 35% of the iron in these foods will be absorbed into the body.[2] Nonheme iron found in plant foods has much lower bioavailability, often 5% or less.[3] Other foods included in the same meal can affect the bioavailability of nonheme iron (see the box "Optimizing Iron Absorption" in chapter 10).[4]

Iron bioavailability in breast milk is between about 15% and 50%, depending on the research methods used to measure it and the babies studied. Babies that are exclusively breastfed or iron-deficient are more efficient at absorbing iron from breast milk; those that are receiving an iron supplement or solid foods are less efficient.[5] The bioavailability of the iron added to infant formula is about 10%.[6]

To return to the WHO's estimated iron requirement, we have to consider bioavailability in the total diet to estimate how much dietary iron is needed. For example, if the average bioavailability of a baby's diet is about 10%, she'll have to consume a total of 9.3 mg of food iron to meet the goal of 0.93 mg of absorbed iron. If she gets most of her iron from plant foods, she'll need much more. If she eats lots of meat and breast milk, she'll need less.

When you understand how the iron requirement is calculated, you can see that, because it is meant to cover the majority of the population, it is intentionally an overestimate for most babies. A smaller-than-average baby or one growing at a slower rate might need much less iron. This explains why some babies do fine with exclusive breastfeeding beyond 6 months and not much in the way of dietary iron, but what works for one baby might be inadequate for another.

APPENDIX H

Should We Be Concerned About Arsenic in Rice Cereal?

Rice cereal has been a popular first infant food for several generations, largely because it is thought to be easily digested and less likely to cause an allergic reaction than other grains. However, we now know that there is a major downside to rice cereal: it can contain high levels of inorganic arsenic. Arsenic is a naturally occurring element that is present in the earth's crust, and thus in our soils and groundwater, but we've added to the arsenic in our environment through mining, fuel burning, and pesticide use. Because it grows in water, rice tends to accumulate a lot of arsenic. After analyzing 69 samples of infant rice cereal, the FDA reported an average of 1.8 μg (micrograms) of inorganic arsenic per serving.[1] Levels in brown rice cereal were generally higher than those in white rice cereal, and it made no difference whether or not the cereals were certified as organically grown. Similar levels of arsenic were found in regular white and brown rice, both organic and conventional, so making your own rice cereal for your baby wouldn't reduce arsenic exposure.

As is true for all chemicals, the poison in arsenic is in the dose. A little bit of arsenic is a part of life; it's in our food and water, after all. How much is too much for a baby to eat every day? We don't have great data on this, particularly for babies, but the EPA estimates that chronic inorganic arsenic intake of more than 0.3 μg/kg per day may cause skin or vascular problems.[2] For a 15-pound baby, that comes out to a limit of 2 μg per day. Based on the FDA's data on rice, just one serving of rice can put a baby close to the limit, and this doesn't include the additional arsenic that might be found in other foods or drinking water.[3]

The EPA's limit for inorganic arsenic intake is based on preventing the skin lesions that occur with arsenic toxicity, but this is just the first symptom of exposure. Chronic exposure to arsenic has also been associated with cancer, neurotoxicity, developmental effects, heart disease, and diabetes.[4] The EPA is in the process of updating its toxicological assessment of inorganic arsenic, so hopefully we'll have more information in the near future.[5] In the meantime, particularly for growing babies, it seems smart to limit their daily exposure to arsenic. An occasional bit of rice cereal or other rice products (like rice cakes and crackers) is probably fine, but for most of the time, I recommend choosing other cereals, like oatmeal, barley, and wheat.

The story of arsenic in rice made a big splash in the media. It's always scary when we learn that a food, especially a staple like rice, may have worrisome levels of a potentially dangerous compound. But it illustrates some of the most important nutrition advice for everyone, babies included: dietary diversity is important. For example, think of families who have chosen to go gluten-free, not out of medical necessity, but because they believe it will help them lose weight or is inherently healthier. (Neither is true.) Many people who avoid gluten instead eat lots of rice. When the arsenic story broke, what was perceived to be the healthier diet now appeared to be laced with a dangerous carcinogen. Dietary diversity protects you against problems like this. Serve rice occasionally, but also explore other interesting grains. There may be unknown risks to other foods, but if you stick to the variety rule, you spread out your risk so that you're never eating too much of any one thing.

NOTES

CHAPTER 1. SHOW ME THE SCIENCE

1. A. J. Wakefield et al., "Ileal-Lymphoid-Nodular Hyperplasia, Non-Specific Colitis, and Pervasive Developmental Disorder in Children," *Lancet* 351, no. 9103 (1998): 637–41.
2. OCEBM Levels of Evidence Working Group, "The Oxford 2011 Levels of Evidence," Oxford Centre for Evidence-Based Medicine, 2011, http://www.cebm.net/index.aspx?o=5653; Yale University Library, "Evidence-Based Clinical Practice Resources," *Library Subject Guides*, accessed June 24, 2014, http://guides.library.yale.edu/content.php?pid=9786&sid=73113; D. Coggon, G. Rose, and D. J. P. Barker, "Epidemiology for the Uninitiated," *BMJ*, accessed June 24, 2014, http://www.bmj.com/about-bmj/resources-readers/publications/epidemiology-uninitiated.
3. M. S. Kramer and R. Kakuma, "Optimal Duration of Exclusive Breastfeeding," *Cochrane Database of Systematic Reviews* 8 (2012): CD003517, doi:10.1002/14651858.CD003517.pub2.
4. O. H. Jonsdottir et al., "Exclusive Breastfeeding for 4 versus 6 Months and Growth in Early Childhood," *Acta Paediatrica* 103, no. 1 (2013): 105–11, doi:10.1111/apa.12433.
5. C. Doctorow, "Correlation between Autism Diagnosis and Organic Food Sales," *Boing Boing* (blog), January 1, 2013, http://boingboing.net/2013/01/01/correlation-between-autism-dia.html.
6. G. S. Fisch, "Nosology and Epidemiology in Autism: Classification Counts," *American Journal of Medical Genetics: Part C, Seminars in Medical Genetics* 160C, no. 2 (2012): 91–103, doi:10.1002/ajmg.c.31325; K. M. Keyes et al., "Cohort Effects Explain the Increase in Autism Diagnosis among Children Born from 1992 to 2003 in California," *International Journal of Epidemiology* 41, no. 2 (2012): 495–503, doi:10.1093/ije/dyr193.

7. R. Gilbert, "Infant Sleeping Position and the Sudden Infant Death Syndrome: Systematic Review of Observational Studies and Historical Review of Recommendations from 1940 to 2002," *International Journal of Epidemiology* 34, no. 4 (2005): 874–87, doi:10.1093/ije/dyi088.
8. P. Franco et al., "Arousal from Sleep Mechanisms in Infants," *Sleep Medicine* 11, no. 7 (2010): 603–14, doi:10.1016/j.sleep.2009.12.014.

CHAPTER 2. CUTTING THE UMBILICAL CORD

1. A. C. Yao, M. Moinian, and J. Lind, "Distribution of Blood between Infant and Placenta after Birth," *Lancet* 294, no. 7626 (1969): 871–73.
2. K. G. Dewey and C. M. Chaparro, "Session 4: Mineral Metabolism and Body Composition Iron Status of Breast-Fed Infants," *Proceedings of the Nutrition Society* 66 (2007): 412–22, doi:10.1017/S002966510700568X; C. M. Chaparro, "Timing of Umbilical Cord Clamping: Effect on Iron Endowment of the Newborn and Later Iron Status," *Nutrition Reviews* 69, suppl. 1 (2011): S30-36, doi:10.1111/j.1753–4887.2011.00430.x.
3. D. A. Erickson-Owens, J. S. Mercer, and W. Oh, "Umbilical Cord Milking in Term Infants Delivered by Cesarean Section: A Randomized Controlled Trial," *Journal of Perinatology* 32, no. 8 (2012): 580–84; J. S. Mercer and D. A. Erickson-Owens, "Rethinking Placental Transfusion and Cord Clamping Issues," *Journal of Perinatal and Neonatal Nursing* 26, no. 3 (2012): 202–17, doi:10.1097/JPN.0b013e31825d2d9a.
4. T. Raju, "Timing of Umbilical Cord Clamping after Birth for Optimizing Placental Transfusion," *Current Opinion in Pediatrics* 25, no. 2 (2013): 180–87, doi:10.1097/MOP.0b013e32835d2a9e.
5. P. M. Dunn, "Dr Erasmus Darwin (1731–1802) of Lichfield and Placental Respiration," *Archives of Disease in Childhood: Fetal and Neonatal Edition* 88, no. 4 (2003): F346-48, doi:10.1136/fn.88.4.F346.
6. Chaparro, "Timing of Umbilical Cord Clamping: Effect on Iron Endowment"; Raju, "Timing of Umbilical Cord Clamping after Birth."
7. Chaparro, "Timing of Umbilical Cord Clamping: Effect on Iron Endowment."
8. Camila Chaparro, telephone interview with the author, August 2, 2013.
9. C. M. Chaparro et al., "Effect of Timing of Umbilical Cord Clamping on Iron Status in Mexican Infants: A Randomised Controlled Trial," *Lancet* 367 (2006): 1997–2004, doi:10.1016/S0140-6736(06)68889–2.
10. Raju, "Timing of Umbilical Cord Clamping."
11. Chaparro, "Timing of Umbilical Cord Clamping: Effect on Iron Endowment."
12. R. D. Baker and F. R. Greer, "Diagnosis and Prevention of Iron Deficiency and Iron-Deficiency Anemia in Infants and Young Children (0–3 Years of Age)," *Pediatrics* 126 (2010): 1040–50, doi:10.1542/peds.2010–2576.
13. Chaparro et al., "Effect of Timing of Umbilical Cord Clamping on Iron Status."
14. S. J. McDonald et al., "Effect of Timing of Umbilical Cord Clamping of Term Infants on Maternal and Neonatal Outcomes," *Cochrane Database of System-*

atic Reviews 7 (2013), http://onlinelibrary.wiley.com/doi/10.1002/14651858.CD004074.pub3/full.

15. B. Lozoff et al., "Long-Lasting Neural and Behavioral Effects of Iron Deficiency in Infancy," *Nutrition Reviews* 64, no. 5, pt. 2 (2006): S34-43.
16. J. C. McCann and B. N. Ames, "An Overview of Evidence for a Causal Relation between Iron Deficiency during Development and Deficits in Cognitive or Behavioral Function," *American Journal of Clinical Nutrition* 85, no. 4 (2007): 931–45.
17. C. Algarín et al., "Iron Deficiency Anemia in Infancy: Long-Lasting Effects on Auditory and Visual System Functioning," *Pediatric Research* 53, no. 2 (2003): 217–23, doi:10.1203/01.PDR.0000047657.23156.55; M. Roncagliolo et al., "Evidence of Altered Central Nervous System Development in Infants with Iron Deficiency Anemia at 6 Mo: Delayed Maturation of Auditory Brainstem Responses," *American Journal of Clinical Nutrition* 68, no. 3 (1998): 683–90.
18. B. Lozoff, E. Jimenez, and J. B. Smith, "Double Burden of Iron Deficiency in Infancy and Low Socioeconomic Status: A Longitudinal Analysis of Cognitive Test Scores to Age 19 Years," *Archives of Pediatrics and Adolescent Medicine* 160, no. 11 (2006): 1108–13; B. Lozoff et al., "Poorer Behavioral and Developmental Outcome More than 10 Years after Treatment for Iron Deficiency in Infancy," *Pediatrics* 105 (April 2000): E51; E. L. Congdon et al., "Iron Deficiency in Infancy Is Associated with Altered Neural Correlates of Recognition Memory at 10 Years," *Journal of Pediatrics* 160, no. 6 (2012): 1027–33, doi:10.1016/j.jpeds.2011.12.011.
19. Lozoff, Jimenez, and Smith, "Double Burden of Iron Deficiency in Infancy"; T. Shafir et al., "Iron Deficiency and Infant Motor Development," *Early Human Development* 84, no. 7 (2008): 479–85, doi:10.1016/j.earlhumdev.2007.12.009.
20. Dewey and Chaparro, "Session 4: Mineral Metabolism."
21. Ibid.
22. M. M. Black et al., "Iron Deficiency and Iron-Deficiency Anemia in the First Two Years of Life: Strategies to Prevent Loss of Developmental Potential," *Nutrition Reviews* 69, suppl. 1 (2011): S64-70, doi:10.1111/j.1753–4887.2011.00435.x; C. Cao and K. O. O'Brien, "Pregnancy and Iron Homeostasis: An Update," *Nutrition Reviews* 71, no. 1 (2013): 35–51, doi:10.1111/j.1753–4887.2012.00550.x.
23. Baker and Greer, "Diagnosis and Prevention of Iron Deficiency."
24. American Academy of Pediatrics (AAP) Section on Breastfeeding et al., "Concerns with Early Universal Iron Supplementation of Breastfeeding Infants," *Pediatrics* 127, no. 4 (2011): e1097, doi:10.1542/peds.2011-0201A; O. Hernell and B. Lönnerdal, "Recommendations on Iron Questioned," *Pediatrics* 127, no. 4 (2011): e1099-1101, doi:10.1542/peds.2011-0201C.
25. K. G. Dewey et al., "Iron Supplementation Affects Growth and Morbidity of Breast-Fed Infants: Results of a Randomized Trial in Sweden and Honduras," *Journal of Nutrition* 132 (November 2002): 3249–55.
26. Baker and Greer, "Diagnosis and Prevention of Iron Deficiency"; J. M. Brotanek

et al., "Iron Deficiency in Early Childhood in the United States: Risk Factors and Racial/Ethnic Disparities," *Pediatrics* 120, no. 3 (2007): 568–75, doi:10.1542/peds.2007–0572.

27. Baker and Greer, "Diagnosis and Prevention of Iron Deficiency."
28. C. M. Chaparro et al., "Early Umbilical Cord Clamping Contributes to Elevated Blood Lead Levels among Infants with Higher Lead Exposure," *Journal of Pediatrics* 151 (2007): 506–12, doi:10.1016/j.jpeds.2007.04.056.
29. Centers for Disease Control and Prevention (CDC), "Lead," accessed July 17, 2013, http://www.cdc.gov/nceh/lead.
30. H. Rabe et al., "Effect of Timing of Umbilical Cord Clamping and Other Strategies to Influence Placental Transfusion at Preterm Birth on Maternal and Infant Outcomes," *Cochrane Database of Systematic Reviews* 8 (2012): CD003248, http://onlinelibrary.wiley.com/doi/10.1002/14651858.CD003248.pub3/pdf/standard.
31. J. S. Mercer et al., "Delayed Cord Clamping in Very Preterm Infants Reduces the Incidence of Intraventricular Hemorrhage and Late-Onset Sepsis: A Randomized, Controlled Trial," *Pediatrics* 117 (April 2006): 1235–42, doi:10.1542/peds.2005–1706; Rabe et al., "Effect of Timing of Umbilical Cord Clamping and Other Strategies."
32. O. Baenziger et al., "The Influence of the Timing of Cord Clamping on Postnatal Cerebral Oxygenation in Preterm Neonates: A Randomized, Controlled Trial," *Pediatrics* 119 (2007): 455–59, doi:10.1542/peds.2006–2725.
33. J. S. Mercer et al., "Seven-Month Developmental Outcomes of Very Low Birth Weight Infants Enrolled in a Randomized Controlled Trial of Delayed versus Immediate Cord Clamping," *Journal of Perinatology* 30 (2010): 11–16, doi:10.1038/jp.2009.170.
34. McDonald et al., "Effect of Timing of Umbilical Cord Clamping of Term Infants."
35. Ibid.; E. K. Hutton and E. S. Hassan, "Late vs. Early Clamping of the Umbilical Cord in Full-Term Neonates: Systematic Review and Meta-analysis of Controlled Trials," JAMA 297 (2007): 1241–52, doi:10.1001/jama.297.11.1241.
36. S. J. McDonald, "Management in the Third Stage of Labour" (dissertation, University of Western Australia, 1996).
37. McDonald et al., "Effect of Timing of Umbilical Cord Clamping of Term Infants."
38. Mercer and Erickson-Owens, "Rethinking Placental Transfusion."
39. Ibid.
40. Tonse Raju, telephone interview with the author, February 11, 2014.
41. Hutton and Hassan, "Late vs Early Clamping of the Umbilical Cord"; McDonald et al., "Effect of Timing of Umbilical Cord Clamping of Term Infants."
42. T. N. K. Raju, "Delayed Cord Clamping: Does Gravity Matter?" *Lancet* 384, no. 9939 (July 2014): 213–14, doi:10.1016/S0140-6736(14)60411–6.
43. N. E. Vain et al., "Effect of Gravity on Volume of Placental Transfusion: A

Multicentre, Randomised, Non-Inferiority Trial," *Lancet* 384, no. 9939 (July 2014): 235–40, doi:10.1016/S0140-6736(14)60197-5.

44. American Congress of Obstetricians and Gynecologists (ACOG) Committee on Obstetric Practice, "Committee Opinion No. 543: Timing of Umbilical Cord Clamping after Birth," *Obstetrics and Gynecology* 120, no. 6 (2012): 1522–26; D. Leduc et al., "Active Management of the Third Stage of Labour: Prevention and Treatment of Postpartum Hemorrhage. Society of Obstetricians and Gynaecologists of Canada Clinical Practice Guideline," *Journal of Obstetrics and Gynaecology Canada* 31, no. 10 (2009): 980–93; D. G. Sweet et al., "European Consensus Guidelines on the Management of Neonatal Respiratory Distress Syndrome in Preterm Infants: 2013 Update," *Neonatology* 103, no. 4 (2013): 353–68, doi:10.1159/000349928.
45. World Health Organization, *Guidelines on Basic Newborn Resuscitation* (Geneva: World Health Organization, 2012), http://www.ncbi.nlm.nih.gov/books/NBK137872; J. M. Perlman et al., "Neonatal Resuscitation: 2010 International Consensus on Cardiopulmonary Resuscitation and Emergency Cardiovascular Care Science with Treatment Recommendations," *Pediatrics* 126, no. 5 (2010): e1319-44, doi:10.1542/peds.2010-2972B; Sweet et al., "European Consensus Guidelines."
46. ACOG Committee on Obstetric Practice, "Committee Opinion No. 543"; Leduc et al., "Active Management of the Third Stage of Labour."
47. ACOG Committee on Obstetric Practice, "Committee Opinion No. 543."
48. Jeffrey Ecker, telephone interview with the author, January 31, 2014.
49. Z. Mei et al., "Assessment of Iron Status in US Pregnant Women from the National Health and Nutrition Examination Survey (NHANES), 1999–2006," *American Journal of Clinical Nutrition* 93, no. 6 (2011): 1312–20, doi:10.3945/ajcn.110.007195.
50. O. Andersson et al., "Effect of Delayed versus Early Umbilical Cord Clamping on Neonatal Outcomes and Iron Status at 4 Months: A Randomised Controlled Trial," *BMJ* 343 (2011): d7157, doi:10.1136/bmj.d7157.
51. Chaparro, interview with the author.
52. Mercer and Erickson-Owens, "Rethinking Placental Transfusion"; Raju, "Timing of Umbilical Cord Clamping after Birth."
53. Raju, "Timing of Umbilical Cord Clamping after Birth."
54. Raju, interview with the author.
55. S. Bhatt et al., "Delaying Cord Clamping until Ventilation Onset Improves Cardiovascular Function at Birth in Preterm Lambs," *Journal of Physiology* 591, no. 8 (2013): 2113–26, doi:10.1113/jphysiol.2012.250084.
56. M. R. Thomas et al., "Providing Newborn Resuscitation at the Mother's Bedside: Assessing the Safety, Usability and Acceptability of a Mobile Trolley," *BMC Pediatrics* 14, no. 1 (2014): 135, doi:10.1186/1471–2431–14–135.
57. H. Rabe et al., "Milking Compared with Delayed Cord Clamping to Increase Placental Transfusion in Preterm Neonates: A Randomized Controlled Trial,"

Obstetrics and Gynecology 117, no. 2, pt. 1 (2011): 205–11, doi:10.1097/AOG.0b013e3181fe46ff; Erickson-Owens, Mercer, and Oh, "Umbilical Cord Milking."

58. J. N. Tolosa et al., "Mankind's First Natural Stem Cell Transplant," *Journal of Cellular and Molecular Medicine* 14, no. 3 (2010): 488–95, doi:10.1111/j.1582-4934.2010.01029.x; Mercer and Erickson-Owens, "Rethinking Placental Transfusion."
59. ACOG Committee on Obstetric Practice, ACOG Committee on Genetics, "Committee Opinion No. 399: Umbilical Cord Blood Banking, Committee Opinion," *Obstetrics and Gynecology* 111, no. 2, pt. 1 (2008): 475–77.
60. AAP Committee on Bioethics, "Children as Hematopoietic Stem Cell Donors," *Pediatrics* 125, no. 2 (2010): 392–404, doi:10.1542/peds.2009-3078; ACOG Committee on Obstetric Practice, "Committee Opinion No. 399."
61. AAP Committee on Bioethics, "Children as Hematopoietic Stem Cell Donors."
62. Children's Hospital Oakland Research Institute, "Sibling Donor Cord Blood Program," accessed June 9, 2014, http://www.chori.org/Services/Sibling_Donor_Cord_Blood_Program/indexcord.html.
63. Mercer and Erickson-Owens, "Rethinking Placental Transfusion."
64. ACOG Committee on Obstetric Practice, "Committee Opinion No. 399."

CHAPTER 3. OF INJECTIONS AND EYE GOOP

1. S. Leavitt, "A Baby Story: Olive Eloise Leavitt," *C'est Si Bon* (blog), January 15, 2014, http://www.cestsibonblog.com/2014/01/a-baby-story-olive-eloise-leavitt.html.
2. S. Leavitt, "The Story of a Miracle: The First 24 Hours," *C'est Si Bon* (blog), February 28, 2014, http://www.cestsibonblog.com/2014/02/the-story-of-miracle-first-24-hours.html.
3. S. Leavitt, "The Story of a Miracle: Getting Out of the Woods," *C'est Si Bon* (blog), March 1, 2014, http://www.cestsibonblog.com/2014/03/the-story-of-miracle-getting-out-of.html.
4. Stefani Leavitt, email to the author, April 8, 2014.
5. Centers for Disease Control and Prevention (CDC), "Notes from the Field: Late Vitamin K Deficiency Bleeding in Infants Whose Parents Declined Vitamin K Prophylaxis—Tennessee, 2013," *Morbidity and Mortality Weekly Report (MMWR)* 62, no. 45 (2013): 901–2.
6. M. J. Shearer, "Vitamin K Deficiency Bleeding (VKDB) in Early Infancy," *Blood Reviews* 23, no. 2 (2009): 49–59, doi:10.1016/j.blre.2008.06.001.
7. J. W. Suttie, "Historical Background," in *Vitamin K in Health and Disease* (Boca Raton, FL: CRC Press, 2009), 1–12.
8. H. Dam, "The Antihaemorrhagic Vitamin of the Chick," *Biochemical Journal* 29, no. 6 (1935): 1273–85.
9. R. Zetterström, "H. C. P. Dam (1895–1976) and E. A. Doisy (1893–1986): The Discovery of Antihaemorrhagic Vitamin and Its Impact on Neonatal Health," *Acta Paediatrica* 95, no. 6 (2006): 642–44, doi:10.1080/08035250600719739.

10. F. R. Greer, "Vitamin K the Basics: What's New?" *Early Human Development* 86, suppl. 1 (2010): 43–47, doi:10.1016/j.earlhumdev.2010.01.015.
11. G. Lippi and M. Franchini, "Vitamin K in Neonates: Facts and Myths," *Blood Transfusion* 9, no. 1 (2011): 4–9, doi:10.2450/2010.0034–10.
12. Shearer, "Vitamin K Deficiency Bleeding (VKDB)."
13. C. A. Crowther and D. D. Crosby, "Vitamin K prior to Preterm Birth for Preventing Neonatal Periventricular Haemorrhage," *Cochrane Database of Systematic Reviews* 1 (2010): CD000229; L. Mandelbrot et al., "Placental Transfer of Vitamin K1 and Its Implications in Fetal Hemostasis," *Thrombosis and Haemostasis* 60, no. 1 (1988): 39–43.
14. H. E. Indyk and D. C. Woollard, "Vitamin K in Milk and Infant Formulas: Determination and Distribution of Phylloquinone and Menaquinone-4," *Analyst* 122, no. 5 (1997): 465–69, doi:10.1039/A608221A.
15. S. Kayata et al., "Vitamin K1 and K2 in Infant Human Liver," *Journal of Pediatric Gastroenterology and Nutrition* 8, no. 3 (1989): 304–7.
16. K. Fujita, F. Kakuya, and S. Ito, "Vitamin K1 and K2 Status and Faecal Flora in Breast Fed and Formula Fed 1-Month-Old Infants," *European Journal of Pediatrics* 152, no. 10 (1993): 852–55, doi:10.1007/BF02073386.
17. Shearer, "Vitamin K Deficiency Bleeding (VKDB)."
18. P. M. Loughnan and P. N. McDougall, "Epidemiology of Late Onset Haemorrhagic Disease: A Pooled Data Analysis," *Journal of Paediatrics and Child Health* 29, no. 3 (1993): 177–81, doi:10.1111/j.1440–1754.1993.tb00480.x.
19. American Academy of Pediatrics (AAP), "Policy Statement: Controversies Concerning Vitamin K and the Newborn," *Pediatrics* 112, no. 1 (2003): 191–92.
20. F. R. Greer, "Vitamin K Status of Lactating Mothers and Their Infants," *Acta Paediatrica* 88 (September 1, 1999): 95–103, doi:10.1111/j.1651–2227.1999.tb01308.x.
21. Shearer, "Vitamin K Deficiency Bleeding (VKDB)."
22. Ibid.
23. S. Leavitt, "Our Alpha-1 Kid," *C'est Si Bon* (blog), March 26, 2014, http://www.cestsibonblog.com/2014/03/our-alpha-1-kid.html.
24. P. M. van Hasselt et al., "Vitamin K Deficiency Bleeding in Cholestatic Infants with Alpha-1-Antitrypsin Deficiency," *Archives of Disease in Childhood: Fetal and Neonatal Edition* 94, no. 6 (2009): F456-60, doi:10.1136/adc.2008.148239; P. M. van Hasselt et al., "Prevention of Vitamin K Deficiency Bleeding in Breastfed Infants: Lessons from the Dutch and Danish Biliary Atresia Registries," *Pediatrics* 121, no. 4 (April 1, 2008): e857-63, doi:10.1542/peds.2007–1788.
25. R. Schulte et al., "Rise in Late Onset Vitamin K Deficiency Bleeding in Young Infants Due to Omission or Refusal of Prophylaxis at Birth," *Pediatric Neurology* 50, no. 6 (2014): 564–68, doi:10.1016/j.pediatrneurol.2014.02.013; R. Decker, "Evidence for the Vitamin K Shot in Newborns," *Evidence Based Birth* (blog), March 18, 2014, http://evidencebasedbirth.com/evidence-for-the-vitamin-k-shot-in-newborns.

26. Schulte et al., "Rise in Late Onset Vitamin K Deficiency Bleeding."
27. J. Golding et al., "Childhood Cancer, Intramuscular Vitamin K, and Pethidine Given during Labour," BMJ 305, no. 6849 (1992): 341–46.
28. Various authors, "Responses to Golding et al. 1992," BMJ, August 8, 1992, http://www.bmj.com/content/305/6849/341.
29. G. J. Draper and C. A. Stiller, "Intramuscular Vitamin K and Childhood Cancer," BMJ 305, no. 6855 (1992): 709–11; R. W. Miller, "Vitamin K and Childhood Cancer," BMJ 305, no. 6860 (1992): 1016.
30. J. A. Ross and S. M. Davies, "Vitamin K Prophylaxis and Childhood Cancer," *Medical and Pediatric Oncology* 34, no. 6 (2000): 434–37, doi:10.1002/(SICI)1096-911X(200006)34:6<434::AID-MPO11>3.0.CO;2-X; E. Roman et al., "Vitamin K and Childhood Cancer: Analysis of Individual Patient Data from Six Case-Control Studies," *British Journal of Cancer* 86, no. 1 (2002): 63–69, doi:10.1038/sj.bjc.6600007; N. T. Fear et al., "Vitamin K and Childhood Cancer: A Report from the United Kingdom Childhood Cancer Study," *British Journal of Cancer* 89, no. 7 (2003): 1228–31, doi:10.1038/sj.bjc.6601278.
31. Hasselt et al., "Prevention of Vitamin K Deficiency Bleeding."
32. B. Laubscher, O. Bänziger, and G. Schubiger, "Prevention of Vitamin K Deficiency Bleeding with Three Oral Mixed Micellar Phylloquinone Doses: Results of a 6-Year (2005–2011) Surveillance in Switzerland," *European Journal of Pediatrics* 172, no. 3 (2013): 357–60, doi:10.1007/s00431-012–1895–1.
33. M. J. Shearer, X. Fu, and S. L. Booth, "Vitamin K Nutrition, Metabolism, and Requirements: Current Concepts and Future Research," *Advances in Nutrition: An International Review Journal* 3, no. 2 (2012): 182–95, doi:10.3945/an.111.001800.
34. Ibid.
35. P. M. Loughnan and P. N. McDougall, "Does Intramuscular Vitamin K1 Act as an Unintended Depot Preparation?" *Journal of Paediatrics and Child Health* 32, no. 3 (1996): 251–54.
36. Stanford School of Medicine, "Guidelines for Vitamin K Prophylaxis," *Newborn Nursery at LPCH*, accessed April 12, 2014, http://newborns.stanford.edu/VitaminK.html.
37. U.S. Food and Drug Administration (FDA), "Q&A on Dietary Supplements," *Dietary Supplements*, March 20, 2014, http://www.fda.gov/Food/DietarySupplements/QADietarySupplements/default.htm#what_info.
38. Crowther and Crosby, "Vitamin K prior to Preterm Birth."
39. B. Pietschnig et al., "Vitamin K in Breast Milk: No Influence of Maternal Dietary Intake," *European Journal of Clinical Nutrition* 47, no. 3 (1993): 209–15.
40. F. R. Greer et al., "Improving the Vitamin K Status of Breastfeeding Infants with Maternal Vitamin K Supplements," *Pediatrics* 99, no. 1 (1997): 88–92, doi:10.1542/peds.99.1.88.
41. U.S. Department of Agriculture (USDA), "Kale, Raw," *National Nutrient Database for Standard Reference*, accessed April 10, 2014, http://ndb.nal.usda.gov/ndb/

foods/show/3030?fg=&man=&lfacet=&format=&count=&max=25&offset=&sort=&qlookup=kale.

42. P. S. Shah et al., "Breastfeeding or Breast Milk for Procedural Pain in Neonates," *Cochrane Database of Systematic Reviews* 12 (2012): CD004950; C. Johnston et al., "Skin-to-Skin Care for Procedural Pain in Neonates," *Cochrane Database of Systematic Reviews* 1 (2014): CD008435.
43. E. Koklu et al., "Anaphylactic Shock due to Vitamin K in a Newborn and Review of Literature," *Journal of Maternal-Fetal and Neonatal Medicine* 27, no. 11 (2014): 1180–81, doi:10.3109/14767058.2013.847425.
44. Frank R. Greer, telephone interview with the author, April 16, 2014.
45. S. J. Fomon, "Infant Feeding and Evolution," in *Nutrition of Normal Infants* (St. Louis, MO: Mosby–Year Book, 1993), 1–5.
46. C. S. F. Credé, "Prevention of Inflammatory Eye Disease in the Newborn," *Archiv für Gynaekologie* 17 (1881): 50–53.
47. J. D. Oriel, "Eminent Venereologists 5: Carl Credé," *Genitourinary Medicine* 67, no. 1 (1991): 67–69; P. M. Dunn, "Dr Carl Credé (1819–1892) and the Prevention of Ophthalmia Neonatorum," *Archives of Disease in Childhood: Fetal and Neonatal Edition* 83, no. 2 (2000): F158-59, doi:10.1136/fn.83.2.F158; Credé, "Prevention of Inflammatory Eye Disease."
48. Credé, "Prevention of Inflammatory Eye Disease."
49. Oriel, "Eminent Venereologists 5"; Dunn, "Dr Carl Credé (1819–1892)"; Credé, "Prevention of Inflammatory Eye Disease."
50. U. C. Schaller and V. Klauss, "Is Credé's Prophylaxis for Ophthalmia Neonatorum Still Valid?" *Bulletin of the World Health Organization* 79, no. 3 (2001): 262–66; E. K. Darling and H. McDonald, "A Meta-analysis of the Efficacy of Ocular Prophylactic Agents Used for the Prevention of Gonococcal and Chlamydial Ophthalmia Neonatorum," *Journal of Midwifery and Women's Health* 55, no. 4 (2010): 319–27, doi:10.1016/j.jmwh.2009.09.003.
51. AAP, "Prevention of Neonatal Ophthalmia," *Red Book Online*, April 14, 2014, http://aapredbook.aappublications.org/content/1/SEC317/SEC329.body.
52. Centers for Disease Control and Prevention (CDC), "Gonococcal Infections: 2010 STD Treatment Guidelines," *Sexually Transmitted Diseases*, 2010, http://www.cdc.gov/std/treatment/2010/gonococcal-infections.htm; I. Mabry-Hernandez and R. Oliverio-Hoffman, "Ocular Prophylaxis for Gonococcal Ophthalmia Neonatorum: Evidence Update for the U.S. Preventive Services Task Force Reaffirmation Recommendation Statement," U.S. Preventive Services Task Force, 2010, http://www.uspreventiveservicestaskforce.org/uspstf10/gonoculproph/gonocup.htm.
53. Zuppa et al., "Ophthalmia Neonatorum: What Kind of Prophylaxis?" *Journal of Maternal-Fetal and Neonatal Medicine* 24, no. 6 (2011): 769–73, doi:10.3109/14767058.2010.531326.
54. Ibid.; Darling and McDonald, "A Meta-analysis of the Efficacy of Ocular Prophylactic Agents."

55. W. J. Benevento et al., "The Sensitivity of *Neisseria gonorrhoeae, Chlamydia trachomatis*, and Herpes Simplex Type II to Disinfection with Povidone-Iodine," *American Journal of Ophthalmology* 109, no. 3 (1990): 329–33; S. J. Isenberg, L. Apt, and M. Wood, "A Controlled Trial of Povidone-Iodine as Prophylaxis against Ophthalmia Neonatorum," *New England Journal of Medicine* 332, no. 9 (1995): 562–66, doi:10.1056/NEJM199503023320903; Z. Ali et al., "Prophylaxis of Ophthalmia Neonatorum: Comparison of Betadine, Erythromycin and No Prophylaxis," *Journal of Tropical Pediatrics* 53, no. 6 (2007): 388–92, doi:10.1093/tropej/fmm049.
56. M. David, S. Rumelt, and Z. Weintraub, "Efficacy Comparison between Povidone Iodine 2.5% and Tetracycline 1% in Prevention of Ophthalmia Neonatorum," *Ophthalmology* 118, no. 7 (2011): 1454–58, doi:10.1016/j.ophtha.2010.12.003.
57. Zuppa et al., "Ophthalmia Neonatorum."
58. World Health Organization (WHO), "Guidelines for the Management of Sexually Transmitted Infections," 2003, http://www.who.int/hiv/pub/sti/pub6/en/; CDC, "Gonococcal Infections"; AAP, "Prevention of Neonatal Ophthalmia."
59. CDC, *Sexually Transmitted Disease Surveillance 2012* (Atlanta: U.S. Department of Health and Human Services, 2014), http://www.cdc.gov/Std/stats12/default.htm; H. Weinstock, S. Berman, and W. Cates, "Sexually Transmitted Diseases among American Youth: Incidence and Prevalence Estimates, 2000," *Perspectives on Sexual and Reproductive Health* 36, no. 1 (2004): 6–10, doi:10.1363/3600604.
60. CDC, "Detailed STD Facts: Gonorrhea," *Sexually Transmitted Diseases*, 2014, http://www.cdc.gov/std/gonorrhea/STDFact-gonorrhea-detailed.htm.
61. CDC, "Recommendations for the Laboratory-Based Detection of *Chlamydia trachomatis* and *Neisseria gonorrhoeae*—2014," MMWR *Recommendations and Reports* 63, no. RR02 (March 14, 2014): 1–19.
62. CDC, "Gonococcal Infections."
63. Zuppa et al., "Ophthalmia Neonatorum."
64. B. Diener, "Cesarean Section Complicated by Gonococcal Ophthalmia Neonatorum," *Journal of Family Practice* 13, no. 5 (1981): 739, 743–44; C. L. Strand and V. A. Arango, "Gonococcal Ophthalmia Neonatorum after Delivery by Cesarean Section: Report of a Case," *Sexually Transmitted Diseases* 6, no. 2 (1979): 77–78.
65. Isenberg, Apt, and Wood, "A Controlled Trial of Povidone-Iodine"; T. A. Bell et al., "Randomized Trial of Silver Nitrate, Erythromycin, and No Eye Prophylaxis for the Prevention of Conjunctivitis among Newborns Not at Risk for Gonococcal Ophthalmitis," *Pediatrics* 92, no. 6 (1993): 755.
66. CDC, "Gonococcal Infections"; A. Matejcek and R. D. Goldman, "Treatment and Prevention of Ophthalmia Neonatorum," *Canadian Family Physician* 59, no. 11 (2013): 1187–90.
67. AAP, "Gonococcal Infections," *Red Book Online*, April 14, 2014, http://aapredbook

.aappublications.org/content/1/SEC131/SEC184.body#graphic-1061; Mabry-Hernandez and Oliverio-Hoffman, "Ocular Prophylaxis."

68. Kristi Watterberg, telephone interview with the author, April 24, 2014.
69. Schaller and Klauss, "Is Credé's Prophylaxis for Ophthalmia Neonatorum Still Valid?"
70. European Centre for Disease Prevention and Control (ECDC), *Sexually Transmitted Infections in Europe 1990–2010* (Stockholm: ECDC, 2012).
71. CDC, *Sexually Transmitted Disease Surveillance 2012*.
72. AAP, "Prevention of Neonatal Ophthalmia."
73. Margaret R. Hammerschlag, email to the author, April 29, 2014.
74. D. Leonard, email to the author, May 1, 2014.
75. P. M. Butterfield, R. N. Emde, and B. B. Platt, "Effects of Silver Nitrate on Initial Visual Behavior," *American Journal of Diseases of Children* 132, no. 4 (1978): 426; P. M. Butterfield, R. N. Emde, and M. J. Svejda, "Does the Early Application of Silver Nitrate Impair Maternal Attachment?," *Pediatrics* 67, no. 5 (1981): 737.
76. AAP Committee on Drugs, Committee on Fetus and Newborn, and Committee on Infectious Diseases, "Prophylaxis and Treatment of Neonatal Gonococcal Infections," *Pediatrics* 65, no. 5 (1980): 1047–48.
77. K. Hedberg et al., "Outbreak of Erythromycin-Resistant Staphylococcal Conjunctivitis in a Newborn Nursery," *Pediatric Infectious Disease Journal* 9, no. 4 (1990): 268–73.
78. M. Fallani et al., "Intestinal Microbiota of 6-Week-Old Infants across Europe: Geographic Influence beyond Delivery Mode, Breast-Feeding, and Antibiotics," *Journal of Pediatric Gastroenterology* 51, no. 1 (2010): 77–84, doi:10.1097/MPG.0b013e3181d1b11e; S. Tanaka et al., "Influence of Antibiotic Exposure in the Early Postnatal Period on the Development of Intestinal Microbiota," *FEMS Immunology and Medical Microbiology* 56, no. 1 (2009): 80–87, doi:10.1111/j.1574-695X.2009.00553.x; S. Matamoros et al., "Development of Intestinal Microbiota in Infants and Its Impact on Health," *Trends in Microbiology* 21, no. 4 (2013): 167–73, doi:10.1016/j.tim.2012.12.001.
79. C. Gray, "Systemic Toxicity with Topical Ophthalmic Medications in Children," *Paediatric and Perinatal Drug Therapy* 7, no. 1 (2006): 23–29.
80. F. Moussa, B. Alaswad, and J. Garcia, "Erythromycin Eye Ointment: Effect on Gastrointestinal Motility," *American Journal of Gastroenterology* 95, no. 3 (2000): 826, doi:10.1111/j.1572–0241.2000.01893.x.
81. S. R. Jadcherla and C. L. Berseth, "Effect of Erythromycin on Gastroduodenal Contractile Activity in Developing Neonates," *Journal of Pediatric Gastroenterology* 34, no. 1 (2002): 16–22.
82. Henry Redel, email to the author, April 24, 2014.
83. R. Decker, "Erythromycin Eye Ointment for Newborns," *Evidence Based Birth* (blog), November 11, 2012, http://evidencebasedbirth.com/is-erythromycin-eye-ointment-always-necessary-for-newborns.

84. A. R. Kemper, "Newborn Screening," *UpToDate*, accessed June 10, 2014, www.uptodate.com.

CHAPTER 4. FOR ONCE, SIT BACK AND WATCH

1. M. Small, *Our Babies, Ourselves: How Biology and Culture Shape the Way We Parent* (New York: Anchor Books, 1999); J. J. McKenna and T. McDade, "Why Babies Should Never Sleep Alone: A Review of the Co-Sleeping Controversy in Relation to SIDS, Bedsharing and Breast Feeding," *Paediatric Respiratory Reviews* 6 (2005): 134–52, doi:10.1016/j.prrv.2005.03.006.
2. Leah R., email to the author, June 9, 2013.
3. A.-M. Widström et al., "Newborn Behaviour to Locate the Breast When Skin-to-Skin: A Possible Method for Enabling Early Self-Regulation," *Acta Paediatrica* 100, no. 1 (2011): 79–85, doi:10.1111/j.1651–2227.2010.01983.x.
4. Ibid.
5. Kym S., telephone interview with the author, June 4, 2013.
6. Widström et al., "Newborn Behaviour to Locate the Breast."
7. Ibid.
8. Mother and Child Health Education Trust, "Breast Crawl: Initiation of Breastfeeding, Emotional Development, Mother-Infant Attachment, Bonding," 2014, http://www.breastcrawl.org.
9. K. A. Bard, "Parenting in Primates," in *Handbook of Parenting, Volume 2: Biology and Ecology of Parenting*, ed. M. H. Bornstein (Mahwah, NJ: Erlbaum, 2008).
10. M. Shibata et al., "Broad Cortical Activation in Response to Tactile Stimulation in Newborns," *Neuroreport* 23, no. 6 (2012): 373–77, doi:10.1097/WNR.0b013e3283520296.
11. S. Kotagal, "Neurological Examination of the Newborn," *UpToDate*, January 10, 2013, http://www.uptodate.com.
12. Stanford Children's Health, "Newborn Reflexes," accessed November 14, 2014, http://www.stanfordchildrens.org/en/topic/default?id=newborn-reflexes-90-P02630.
13. P. Rochat, "Five Levels of Self-Awareness as They Unfold Early in Life," *Consciousness and Cognition* 12, no. 4 (2003): 717–31, doi:10.1016/S1053-8100(03)00081–3.
14. A. Streri, M. Lhote, and S. Dutilleul, "Haptic Perception in Newborns," *Developmental Science* 3, no. 3 (2000): 319–27.
15. L. Marcus et al., "Tactile Sensory Capacity of the Preterm Infant: Manual Perception of Shape from 28 Gestational Weeks," *Pediatrics* 130, no. 1 (2012): e88-94, doi:10.1542/peds.2011–3357.
16. P. Rochat, "Mouthing and Grasping in Neonates: Evidence for the Early Detection of What Hard or Soft Substances Afford for Action," *Infant Behavior and Development* 10, no. 4 (1987): 435–49, doi:10.1016/0163–6383(87)90041–5.

17. L. Gray, L. Watt, and E. M. Blass, "Skin-to-Skin Contact Is Analgesic in Healthy Newborns," *Pediatrics* 105, no. 1 (2000): e14.
18. B. S. Kisilevsky et al., "Effects of Experience on Fetal Voice Recognition," *Psychological Science* 14, no. 3 (2003): 220–24.
19. C. Moon, R. Panneton Cooper, and W. P. Fifer, "Two-Day-Olds Prefer Their Native Language," *Infant Behavior and Development* 16, no. 4 (1993): 495–500, doi:10.1016/0163–6383(93)80007-U.
20. A. J. DeCasper et al., "Fetal Reactions to Recurrent Maternal Speech," *Infant Behavior and Development* 17, no. 2 (1994): 159–64.
21. A. J. DeCasper and W. P. Fifer, "Of Human Bonding: Newborns Prefer Their Mothers' Voices," *Science* 208, no. 4448 (1980): 1174–76.
22. H. Kurihara et al., "Behavioral and Adrenocortical Responses to Stress in Neonates and the Stabilizing Effects of Maternal Heartbeat on Them," *Early Human Development* 46, no. 1–2 (1996): 117–27, doi:10.1016/0378–3782(96)01749–5.
23. P. Glass, "Development of the Visual System and Implications for Early Intervention," *Infants and Young Children* 15, no. 1 (2002): 1–10.
24. D. Y. Teller, "First Glances: The Vision of Infants. The Friedenwald Lecture," *Investigative Ophthalmology and Visual Science* 38, no. 11 (1997): 2183–203.
25. Y. Pan et al., "Visual Acuity Norms in Preschool Children: The Multi-Ethnic Pediatric Eye Disease Study," *Optometry and Vision Science: Official Publication of the American Academy of Optometry* 86, no. 6 (2009): 607–12, doi:10.1097/OPX.0b013e3181a76e55.
26. R. G. Bosworth and K. R. Dobkins, "Chromatic and Luminance Contrast Sensitivity in Fullterm and Preterm Infants," *Journal of Vision* 9, no. 13 (2009): 15, doi:10.1167/9.13.15.
27. D. Ricci et al., "Application of a Neonatal Assessment of Visual Function in a Population of Low Risk Full-Term Newborn," *Early Human Development* 84, no. 4 (2008): 277–80, doi:10.1016/j.earlhumdev.2007.10.002; Glass, "Development of the Visual System"; L. M. Dubowitz et al., "Visual Function in the Preterm and Fullterm Newborn Infant," *Developmental Medicine and Child Neurology* 22, no. 4 (1980): 465–75.
28. C. C. Goren, M. Sarty, and P. Y. K. Wu, "Visual Following and Pattern Discrimination of Face-Like Stimuli by Newborn Infants," *Pediatrics* 56, no. 4 (1975): 544–49.
29. O. R. Salva et al., "The Evolution of Social Orienting: Evidence from Chicks (*Gallus gallus*) and Human Newborns," *PLoS ONE* 6, no. 4 (2011): e18802, doi:10.1371/journal.pone.0018802; E. Valenza et al., "Face Preference at Birth," *Journal of Experimental Psychology: Human Perception and Performance* 22, no. 4 (1996): 892–903.
30. A. Batki et al., "Is There an Innate Gaze Module? Evidence from Human Neonates," *Infant Behavior and Development* 23, no. 2 (2000): 223–29, doi:10.1016/S0163-6383(01)00037–6.
31. T. Farroni et al., "Eye Contact Detection in Humans from Birth," *Proceedings*

of the National Academy of Sciences of the United States of America 99, no. 14 (2002): 9602–5, doi:10.1073/pnas.152159999.

32. F. Z. Sai, "The Role of the Mother's Voice in Developing Mother's Face Preference: Evidence for Intermodal Perception at Birth," *Infant and Child Development* 14, no. 1 (2005): 29–50, doi:10.1002/icd.376.
33. S. Brink, *The Fourth Trimester: Understanding, Protecting, and Nurturing an Infant through the First Three Months* (Berkeley, CA: University of California Press, 2013).
34. Widström et al., "Newborn Behaviour to Locate the Breast."
35. Penny Glass, email to the author, June 25, 2013.
36. H. Varendi, R. H. Porter, and J. Winberg, "Does the Newborn Baby Find the Nipple by Smell?" *Lancet* 344 (1994): 989–90.
37. H. Varendi, R. H. Porter, and J. Winberg, "Natural Odour Preferences of Newborn Infants Change over Time," *Acta Paediatrica* 86, no. 9 (1997): 985–90.
38. H. Varendi, R. H. Porter, and J. Winberg, "Attractiveness of Amniotic Fluid Odor: Evidence of Prenatal Olfactory Learning?" *Acta Paediatrica* 85, no. 10 (1996): 1223–27.
39. H. Varendi et al., "Soothing Effect of Amniotic Fluid Smell in Newborn Infants," *Early Human Development* 51, no. 1 (1998): 47–55, doi:10.1016/S0378-3782(97)00082-0.
40. Varendi, Porter, and Winberg, "Natural Odour Preferences."
41. R. H. Porter and J. Winberg, "Unique Salience of Maternal Breast Odors for Newborn Infants," *Neuroscience and Biobehavioral Reviews* 23, no. 3 (1999): 439–49.
42. J. M. Cernoch and R. H. Porter, "Recognition of Maternal Axillary Odors by Infants," *Child Development* 56 (1985): 1593–98.
43. J. A. Mennella, A. Johnson, and G. K. Beauchamp, "Garlic Ingestion by Pregnant Women Alters the Odor of Amniotic Fluid," *Chemical Senses* 20, no. 2 (1995): 207–9.
44. J. A. Mennella and G. K. Beauchamp, "Maternal Diet Alters the Sensory Qualities of Human Milk and the Nursling's Behavior," *Pediatrics* 88 (1991): 737–44.
45. MedlinePlus, National Library of Medicine, "APGAR," *MedlinePlus Medical Encyclopedia*, accessed January 14, 2014, http://www.nlm.nih.gov/medlineplus/ency/article/003402.htm.
46. H. Lagercrantz, "Stress, Arousal, and Gene Activation at Birth," *Physiology* 11, no. 5 (1996): 214–18.
47. O. Romantshik et al., "Preliminary Evidence of a Sensitive Period for Olfactory Learning by Human Newborns," *Acta Paediatrica* 96, no. 3 (2007): 372–76, doi:10.1111/j.1651-2227.2006.00106.x.
48. H. Varendi, R. H. Porter, and J. Winberg, "The Effect of Labor on Olfactory Exposure Learning within the First Postnatal Hour," *Behavioral Neuroscience* 116, no. 2 (2002): 206–11.

49. R. M. Sullivan et al., “Association of an Odor with Activation of Olfactory Bulb Noradrenergic Beta-Receptors or Locus Coeruleus Stimulation Is Sufficient to Produce Learned Approach Responses to That Odor in Neonatal Rats,” *Behavioral Neuroscience* 114, no. 5 (2000): 957–62.
50. M. Kaitz and R. Bronner, “Parturient Women Can Recognize Their Infants by Touch,” *Developmental Psychology* 28, no. 1 (1992): 35–39.
51. M. Kaitz et al., “Mothers’ and Fathers’ Recognition of Their Newborns’ Photographs during the Postpartum Period,” *Developmental and Behavioral Pediatrics* 9, no. 4 (1988): 223–26.
52. J. N. Lundström et al., “Maternal Status Regulates Cortical Responses to the Body Odor of Newborns,” *Frontiers in Psychology* 4 (2013), doi:10.3389/fpsyg.2013.00597.
53. M. Kaitz et al., “Mothers’ Recognition of Their Newborns by Olfactory Cues,” *Developmental Psychobiology* 20, no. 6 (1987): 587–91.
54. R. C. White-Traut et al., “Salivary Cortisol and Behavioral State Responses of Healthy Newborn Infants to Tactile-Only and Multisensory Interventions,” *Journal of Obstetric, Gynecologic, and Neonatal Nursing* 38, no. 1 (2009): 22–34, doi:10.1111/j.1552–6909.2008.00307.x.
55. C. Blair et al., “Maternal Sensitivity Is Related to Hypothalamic-Pituitary-Adrenal Axis Stress Reactivity and Regulation in Response to Emotion Challenge in 6-Month-Old Infants,” *Annals of the New York Academy of Sciences* 1094, no. 1 (2006): 263–67, doi:10.1196/annals.1376.031.
56. E. M. Albers et al., “Maternal Behavior Predicts Infant Cortisol Recovery from a Mild Everyday Stressor,” *Journal of Child Psychology and Psychiatry* 49, no. 1 (2008): 97–103, doi:10.1111/j.1469–7610.2007.01818.x.
57. J. R. Britton, H. L. Britton, and V. Gronwaldt, “Breastfeeding, Sensitivity, and Attachment,” *Pediatrics* 118, no. 5 (2006): e1436-43, doi:10.1542/peds.2005–2916; A. Tharner et al., “Breastfeeding and Its Relation to Maternal Sensitivity and Infant Attachment,” *Journal of Developmental and Behavioral Pediatrics* 33, no. 5 (2012): 396–404; Douglas M. Teti, telephone interview with the author, February 11, 2014.
58. N. Koren-Karie et al., “Mothers’ Insightfulness Regarding Their Infants’ Internal Experience: Relations with Maternal Sensitivity and Infant Attachment,” *Developmental Psychology* 38, no. 4 (July 2002): 534–42, doi:10.1037/0012–1649.38.4.534; N. L. McElwain and C. Booth-LaForce, “Maternal Sensitivity to Infant Distress and Nondistress as Predictors of Infant-Mother Attachment Security,” *Journal of Family Psychology* 20, no. 2 (2006): 247–55, doi:10.1037/0893–3200.20.2.247; E. M. Leerkes, A. Nayena Blankson, and M. O’Brien, “Differential Effects of Maternal Sensitivity to Infant Distress and Nondistress on Social-Emotional Functioning,” *Child Development* 80, no. 3 (2009): 762–75; S. H. Landry, K. E. Smith, and P. R. Swank, “Responsive Parenting: Establishing Early Foundations for Social, Communication, and Independent Problem-Solving Skills,” *Developmental Psychology* 42, no. 4 (2006): 627–42, doi:10.1037/0012–1649.42.4.627.

CHAPTER 5. MILK AND MOTHERHOOD

1. Jordan P. Green, telephone interview with the author, May 25, 2014; Cheryl R. L. Green, telephone interview with the author, May 25, 2014.
2. J. P. Green, interview with the author.
3. C. R. L. Green, interview with the author.
4. S. J. Fomon, "History," in *Nutrition of Normal Infants* (St. Louis, MO: Mosby–Year Book, 1993), 6–14; S. J. Fomon, "Infant Feeding in the 20th Century: Formula and Beikost," *Journal of Nutrition* 131 (2001): 409-20S; D. Thulier, "Breastfeeding in America: A History of Influencing Factors," *Journal of Human Lactation* 25, no. 1 (2009): 85–94, doi:10.1177/0890334408324452.
5. M. Obladen, "Pap, Gruel, and Panada: Early Approaches to Artificial Infant Feeding," *Neonatology* 105, no. 4 (2014): 267–74, doi:10.1159/000357935.
6. Thulier, "Breastfeeding in America."
7. Obladen, "Pap, Gruel, and Panada."
8. Fomon, "History"; Obladen, "Pap, Gruel, and Panada."
9. B. Lozoff, "Birth and Bonding in Non-industrial Societies," *Developmental Medicine and Child Neurology* 25 (1983): 595–600.
10. Fomon, "History."
11. Ibid.
12. Ibid.
13. Ibid.
14. Thulier, "Breastfeeding in America."
15. T. Cassidy, "The Hut, the Home, and the Hospital," in *Birth: The Surprising History of How We Are Born* (New York: Atlantic Monthly Press, 2006), 54–63; E. Temkin, "Rooming-In: Redesigning Hospitals and Motherhood in Cold War America," *Bulletin of the History of Medicine* 76, no. 2 (2002): 271–98, doi:10.1353/bhm.2002.0101.
16. Thulier, "Breastfeeding in America."
17. Fomon, "Infant Feeding in the 20th Century."
18. Thulier, "Breastfeeding in America."
19. F. Hassiotou and D. T. Geddes, "Programming of Appetite Control during Breastfeeding as a Preventative Strategy against the Obesity Epidemic," *Journal of Human Lactation* 30, no. 2 (2014): 136–42, doi:10.1177/0890334414526950; M. Van Den Driessche et al., "Gastric Emptying in Formula-Fed and Breast-Fed Infants Measured with the 13C-Octanoic Acid Breath Test," *Journal of Pediatric Gastroenterology* 29, no. 1 (1999): 46–51.
20. B. Lönnerdal, "Bioactive Proteins in Breast Milk," *Journal of Paediatrics and Child Health* 49 (2013): 1–7, doi:10.1111/jpc.12104; O. Ballard and A. L. Morrow, "Human Milk Composition," *Pediatric Clinics of North America* 60, no. 1 (2013): 49–74, doi:10.1016/j.pcl.2012.10.002.
21. G. Der, G. D. Batty, and I. J. Deary, "Effect of Breast Feeding on Intelligence in

Children: Prospective Study, Sibling Pairs Analysis, and Meta-analysis," *BMJ* 333, no. 7575 (2006): 945, doi:10.1136/bmj.38978.699583.55.

22. Ibid.
23. M. S. Kramer, "Methodological Challenges in Studying Long-Term Effects of Breast-Feeding," in *Breast-Feeding: Early Influences on Later Health*, ed. G. Goldberg et al., Advances in Experimental Medicine and Biology 639 (Dordrecht: Springer Netherlands, 2009), 121–33.
24. S. Ip et al., "Breastfeeding and Maternal and Infant Health Outcomes in Developed Countries," *Evidence Report/Technology Assessment*, no. 153 (2007): 1–186; L. Duijts, M. K. Ramadhani, and H. A. Moll, "Breastfeeding Protects against Infectious Diseases during Infancy in Industrialized Countries: A Systematic Review," *Maternal and Child Nutrition* 5, no. 3 (2009): 199–210, doi:10.1111/j.1740-8709.2008.00176.x; S. W. Abrahams and M. H. Labbok, "Breastfeeding and Otitis Media: A Review of Recent Evidence," *Current Allergy and Asthma Reports* 11, no. 6 (2011): 508–12, doi:10.1007/s11882-011-0218-3.
25. Ip et al., "Breastfeeding and Maternal and Infant Health."
26. Duijts, Ramadhani, and Moll, "Breastfeeding Protects against Infectious Diseases"; L. Duijts et al., "Prolonged and Exclusive Breastfeeding Reduces the Risk of Infectious Diseases in Infancy," *Pediatrics* 126, no. 1 (2010): e18-25, doi:10.1542/peds.2008–3256; M. A. Quigley, Y. J. Kelly, and A. Sacker, "Breastfeeding and Hospitalization for Diarrheal and Respiratory Infection in the United Kingdom Millennium Cohort Study," *Pediatrics* 119, no. 4 (2007): e837-42, doi:10.1542/peds.2006–2256.
27. C. M. Fisk et al., "Breastfeeding and Reported Morbidity during Infancy: Findings from the Southampton Women's Survey," *Maternal and Child Nutrition* 7, no. 1 (2011): 61–70, doi:10.1111/j.1740–8709.2010.00241.x; Quigley, Kelly, and Sacker, "Breastfeeding and Hospitalization"; M. Tarrant et al., "Breast-Feeding and Childhood Hospitalizations for Infections," *Epidemiology* 21, no. 6 (2010): 847–54, doi:10.1097/EDE.0b013e3181f55803.
28. Ballard and Morrow, "Human Milk Composition"; Lönnerdal, "Bioactive Proteins in Breast Milk"; Abrahams and Labbok, "Breastfeeding and Otitis Media."
29. Abrahams and Labbok, "Breastfeeding and Otitis Media."
30. S. B. Tully, Y. Bar-Haim, and R. L. Bradley, "Abnormal Tympanography after Supine Bottle Feeding," *Journal of Pediatrics* 126, no. 6 (1995): S105-11.
31. F. R. Hauck et al., "Breastfeeding and Reduced Risk of Sudden Infant Death Syndrome: A Meta-analysis," *Pediatrics* 128, no. 1 (2011): 103–10, doi:10.1542/peds.2010–3000.
32. J. Blood-Siegfried, "The Role of Infection and Inflammation in Sudden Infant Death Syndrome," *Immunopharmacology and Immunotoxicology* 31, no. 4 (2009): 516–23, doi:10.3109/08923970902814137.
33. R. S. Horne et al., "Comparison of Evoked Arousability in Breast and Formula Fed Infants," *Archives of Disease in Childhood* 89, no. 1 (2004): 22–25.

34. Ip et al., "Breastfeeding and Maternal and Infant Health"; S. Sullivan et al., "An Exclusively Human Milk–Based Diet Is Associated with a Lower Rate of Necrotizing Enterocolitis Than a Diet of Human Milk and Bovine Milk–Based Products," *Journal of Pediatrics* 156, no. 4 (2010): 562–67, doi:10.1016/j.jpeds.2009.10.040.
35. M. L. Kwan et al., "Breastfeeding and the Risk of Childhood Leukemia: A Meta-analysis," *Public Health Reports* 119, no. 6 (2004): 521–35; Ip et al., "Breastfeeding and Maternal and Infant Health"; R. M. Martin et al., "Breast-Feeding and Childhood Cancer: A Systematic Review with Metaanalysis," *International Journal of Cancer* 117, no. 6 (2005): 1020–31, doi:10.1002/ijc.21274.
36. C. G. Colen and D. M. Ramey, "Is Breast Truly Best? Estimating the Effects of Breastfeeding on Long-Term Child Health and Wellbeing in the United States Using Sibling Comparisons," *Social Science and Medicine* 109 (2014): 55–65, doi:10.1016/j.socscimed.2014.01.027.
37. E. Innes, "Breast Milk Is No Better for a Baby Than Bottled Milk, Expert Claims," *Mail Online*, February 26, 2014, http://www.dailymail.co.uk/health/article-2568426/Breast-milk-no-better-baby-bottled-milk-INCREASES-risk-asthma-expert-claims.html; J. Grose, "New Study Confirms It: Breastfeeding Benefits Have Been Drastically Overstated," *Slate*, February 27, 2014, http://www.slate.com/blogs/xx_factor/2014/02/27/breast_feeding_study_benefits_of_breast_over_bottle_have_been_exaggerated.html.
38. T. Cassels, "'Is Breast Really Best?' The Debate Doesn't End Here . . . ," *Evolutionary Parenting*, February 28, 2014, http://evolutionaryparenting.com/is-breast-really-best-the-debate-doesnt-end-here; M. Beyer, "New 'Breast Isn't Best' Study Has the Potential to Derail Nursing," *SheKnows Parenting*, February 28, 2014, http://www.sheknows.com/parenting/articles/1031627/new-breast-isnt-best-study-has-the-potential-to-derail-nursing.
39. American Academy of Pediatrics (AAP) Section on Breastfeeding, "Breastfeeding and the Use of Human Milk," *Pediatrics* 129, no. 3 (2012): e827-41, doi:10.1542/peds.2011–3552.
40. Let's Move!, "Healthy Moms," accessed May 28, 2014, http://www.letsmove.gov/healthy-moms.
41. M. S. Kramer et al., "Promotion of Breastfeeding Intervention Trial (PROBIT): A Randomized Trial in the Republic of Belarus," *JAMA* 285, no. 4 (January 24, 2001): 413–20, doi:10.1001/jama.285.4.413.
42. Ibid.
43. M. S. Kramer, "'Breast Is Best': The Evidence," *Early Human Development* 86, no. 11 (2010): 729–32, doi:10.1016/j.earlhumdev.2010.08.005; M. S. Kramer et al., "Effects of Prolonged and Exclusive Breastfeeding on Child Behavior and Maternal Adjustment: Evidence from a Large, Randomized Trial," *Pediatrics* 121, no. 3 (2008): e435-40, doi:10.1542/peds.2007–1248; M. S. Kramer et al., "Effect of Prolonged and Exclusive Breast Feeding on Risk of Allergy and Asthma: Cluster Randomised Trial," *BMJ* 335, no. 7624 (2007): 815, doi:10.1136/

bmj.39304.464016.AE; M. S. Kramer et al., "A Randomized Breast-Feeding Promotion Intervention Did Not Reduce Child Obesity in Belarus," *Journal of Nutrition* 139, no. 2 (2009): 417-21S, doi:10.3945/jn.108.097675.

44. R. M. Martin et al., "Effects of Promoting Longer-Term and Exclusive Breastfeeding on Cardiometabolic Risk Factors at Age 11.5 Years: A Cluster-Randomized, Controlled Trial," *Circulation* 129, no. 3 (2014): 321–29, doi:10.1161/CIRCULATIONAHA.113.005160; R. M. Martin et al., "Effects of Promoting Longer-Term and Exclusive Breastfeeding on Adiposity and Insulin-like Growth Factor-I at Age 11.5 Years: A Randomized Trial," *JAMA* 309, no. 10 (2013): 1005–13, doi:10.1001/jama.2013.167.
45. M. S. Kramer et al., "Breastfeeding and Child Cognitive Development: New Evidence from a Large Randomized Trial," *Archives of General Psychiatry* 65, no. 5 (2008): 578–84, doi:10.1001/archpsyc.65.5.578.
46. M. B. Belfort et al., "Infant Feeding and Childhood Cognition at Ages 3 and 7 Years: Effects of Breastfeeding Duration and Exclusivity," *JAMA Pediatrics* 167, no. 9 (2013): 836, doi:10.1001/jamapediatrics.2013.455.
47. M. J. Brion et al., "What Are the Causal Effects of Breastfeeding on IQ, Obesity and Blood Pressure? Evidence from Comparing High-Income with Middle-Income Cohorts," *International Journal of Epidemiology* 40, no. 3 (2011): 670–80, doi:10.1093/ije/dyr020.
48. E. Evenhouse and S. Reilly, "Improved Estimates of the Benefits of Breastfeeding Using Sibling Comparisons to Reduce Selection Bias," *Health Services Research* 40, no. 6, pt. 1 (2005): 1781–802, doi:10.1111/j.1475–6773.2005.00453.x; Colen and Ramey, "Is Breast Truly Best?"; Der, Batty, and Deary, "Effect of Breast Feeding on Intelligence."
49. S. M. Innis, "Impact of Maternal Diet on Human Milk Composition and Neurological Development of Infants," *American Journal of Clinical Nutrition* 99, no. 3 (2014): 734-41S, doi:10.3945/ajcn.113.072595.
50. K. F. Michaelsen, L. Lauritzen, and E. L. Mortensen, "Effects of Breast-Feeding on Cognitive Function," in Goldberg et al., *Breast-Feeding*, 199–215.
51. Kramer et al., "A Randomized Breast-Feeding Promotion Intervention"; Martin et al., "Effects of Promoting Longer-Term and Exclusive Breastfeeding on Cardiometabolic Risk"; Martin et al., "Effects of Promoting Longer-Term and Exclusive Breastfeeding on Adiposity."
52. Evenhouse and Reilly, "Improved Estimates of the Benefits of Breastfeeding."
53. C. G. Owen et al., "The Effect of Breastfeeding on Mean Body Mass Index throughout Life: A Quantitative Review of Published and Unpublished Observational Evidence," *American Journal of Clinical Nutrition* 82, no. 6 (December 1, 2005): 1298–1307; C. G. Owen et al., "Effect of Infant Feeding on the Risk of Obesity across the Life Course: A Quantitative Review of Published Evidence," *Pediatrics* 115, no. 5 (2005): 1367–77, doi:10.1542/peds.2004–1176.
54. C. H. Fall et al., "Infant-Feeding Patterns and Cardiovascular Risk Factors in Young Adulthood: Data from Five Cohorts in Low- and Middle-Income

Countries," *International Journal of Epidemiology* 40, no. 1 (2011): 47–62, doi:10.1093/ije/dyq155; Brion et al., "What Are the Causal Effects of Breastfeeding on IQ?"

55. C. M. Dogaru et al., "Breastfeeding and Childhood Asthma: Systematic Review and Meta-analysis," *American Journal of Epidemiology* 179, no. 10 (2014): 1153–67, doi:10.1093/aje/kwu072.
56. M. S. Kramer et al., "Effect of Prolonged and Exclusive Breast Feeding on Risk of Allergy and Asthma: Cluster Randomised Trial," *BMJ* 335, no. 7624 (2007): 815, doi:10.1136/bmj.39304.464016.AE.
57. A. L. Wright et al., "Factors Influencing the Relation of Infant Feeding to Asthma and Recurrent Wheeze in Childhood," *Thorax* 56, no. 3 (2001): 192–97, doi:10.1136/thorax.56.3.192.
58. M. R. Sears et al., "Long-Term Relation between Breastfeeding and Development of Atopy and Asthma in Children and Young Adults: A Longitudinal Study," *Lancet* 360, no. 9337 (2002): 901–7, doi:10.1016/S0140-6736(02)11025–7; M. C. Matheson, K. J. Allen, and M. L. K. Tang, "Understanding the Evidence for and against the Role of Breastfeeding in Allergy Prevention," *Clinical and Experimental Allergy* 42, no. 6 (2012): 827–51, doi:10.1111/j.1365–2222.2011.03925.x.
59. Wright et al., "Factors Influencing the Relation of Infant Feeding to Asthma"; M. S. Kramer, "Invited Commentary: Does Breastfeeding Protect against 'Asthma'?" *American Journal of Epidemiology* 179, no. 10 (2014): 1168–70, doi:10.1093/aje/kwu070.
60. Wright et al., "Factors Influencing the Relation of Infant Feeding to Asthma."
61. C. M. Dieterich et al., "Breastfeeding and Health Outcomes for the Mother-Infant Dyad," *Pediatric Clinics of North America* 60, no. 1 (2013): 31–48, doi:10.1016/j.pcl.2012.09.010; E. Oken et al., "Effects of an Intervention to Promote Breastfeeding on Maternal Adiposity and Blood Pressure at 11.5 Y Postpartum: Results from the Promotion of Breastfeeding Intervention Trial, a Cluster-Randomized Controlled Trial," *American Journal of Clinical Nutrition* 98, no. 4 (2013): 1048–56, doi:10.3945/ajcn.113.065300.
62. AAP Section on Breastfeeding, "Breastfeeding and the Use of Human Milk."
63. F. R. Greer, S. H. Sicherer, and A. W. Burks, "Effects of Early Nutritional Interventions on the Development of Atopic Disease in Infants and Children: The Role of Maternal Dietary Restriction, Breastfeeding, Timing of Introduction of Complementary Foods, and Hydrolyzed Formulas," *Pediatrics* 121, no. 1 (2008): 183–91, doi:10.1542/peds.2007–3022.
64. S. J. Knaak, "The Problem with Breastfeeding Discourse," *Canadian Journal of Public Health* 97, no. 5 (2006): 412–14.
65. AAP Section on Breastfeeding, "Breastfeeding and the Use of Human Milk."
66. Centers for Disease Control and Prevention (CDC), *Breastfeeding Report Card—United States 2013* (Atlanta: CDC, 2013), http://www.cdc.gov/breastfeeding/data/reportcard.htm.

67. A. L. Wright and R. J. Schanler, "The Resurgence of Breastfeeding at the End of the Second Millennium," *Journal of Nutrition* 131, no. 2 (2001): 421-25S.
68. Advisory Panel on the Marketing in Australia of Infant Formula, "An International Comparison Study into the Implementation of the WHO Code and Other Breastfeeding Initiatives," Australian Government Department of Health, 2012, http://www.health.gov.au/internet/publications/publishing.nsf/Content/int-comp-whocode-bf-init~int-comp-whocode-bf-init-ico~int-comp-whocode-bf-init-ico-usa.
69. Senate and House of Representatives of the United States of America in Congress, *Patient Protection and Affordable Care Act of 2010*, H.R. 3590, 2010.
70. Advisory Panel on the Marketing in Australia of Infant Formula, "An International Comparison Study."
71. M. Neifert et al., "The Influence of Breast Surgery, Breast Appearance, and Pregnancy-Induced Breast Changes on Lactation Sufficiency as Measured by Infant Weight Gain," *Birth* 17, no. 1 (1990): 31–38, doi:10.1111/j.1523-536X.1990.tb00007.x.
72. C. J. Chantry et al., "Excess Weight Loss in First-Born Breastfed Newborns Relates to Maternal Intrapartum Fluid Balance," *Pediatrics* 127, no. 1 (2011): e171-79, doi:10.1542/peds.2009–2663.
73. Academy of Breastfeeding Medicine Protocol Committee, "ABM Clinical Protocol #3: Hospital Guidelines for the Use of Supplementary Feedings in the Healthy Term Breastfed Neonate," *Breastfeeding Medicine* 4, no. 3 (2009): 175–82, doi:10.1089/bfm.2009.9991; S. J. Oddie et al., "Severe Neonatal Hypernatraemia: A Population Based Study," *Archives of Disease in Childhood: Fetal and Neonatal Edition* 98, no. 5 (2013): F384-87, doi:10.1136/archdischild-2012–302908; M. L. Moritz et al., "Breastfeeding-Associated Hypernatremia: Are We Missing the Diagnosis?" *Pediatrics* 116, no. 3 (2005): e343-47, doi:10.1542/peds.2004–2647.
74. A. M. Stuebe et al., "Prevalence and Risk Factors for Early, Undesired Weaning Attributed to Lactation Dysfunction," *Journal of Women's Health* 23, no. 5 (2014): 404–12, doi:10.1089/jwh.2013.4506.
75. M. Neifert and M. Bunik, "Overcoming Clinical Barriers to Exclusive Breastfeeding," *Pediatric Clinics of North America* 60, no. 1 (2013): 115–45, doi:10.1016/j.pcl.2012.10.001; R. Grajeda and R. Pérez-Escamilla, "Stress during Labor and Delivery Is Associated with Delayed Onset of Lactation among Urban Guatemalan Women," *Journal of Nutrition* 132, no. 10 (2002): 3055–60; P. Zhu et al., "New Insight into Onset of Lactation: Mediating the Negative Effect of Multiple Perinatal Biopsychosocial Stress on Breastfeeding Duration," *Breastfeeding Medicine* 8 (2013): 151–58, doi:10.1089/bfm.2012.0010.
76. S. L. Matias et al., "Risk Factors for Early Lactation Problems among Peruvian Primiparous Mothers," *Maternal and Child Nutrition* 6, no. 2 (2010): 120–33, doi:10.1111/j.1740–8709.2009.00195.x; G. E. Otoo et al., "HIV-Negative Status Is

Associated with Very Early Onset of Lactation among Ghanaian Women," *Journal of Human Lactation* 26, no. 2 (2010): 107–17, doi:10.1177/0890334409348214.

77. A. M. Stuebe, K. Grewen, and S. Meltzer-Brody, "Association between Maternal Mood and Oxytocin Response to Breastfeeding," *Journal of Women's Health* 22, no. 4 (2013): 352–61, doi:10.1089/jwh.2012.3768.
78. E. Sibolboro Mezzacappa and E. S. Katkin, "Breast-Feeding Is Associated with Reduced Perceived Stress and Negative Mood in Mothers," *Health Psychology* 21, no. 2 (2002): 187–93, doi:10.1037/0278-6133.21.2.187.
79. Stuebe, Grewen, and Meltzer-Brody, "Association between Maternal Mood and Oxytocin Response."
80. A. M. Stuebe et al., "Failed Lactation and Perinatal Depression: Common Problems with Shared Neuroendocrine Mechanisms?" *Journal of Women's Health* 21, no. 3 (2012): 264–72, doi:10.1089/jwh.2011.3083.
81. Stuebe et al., "Prevalence and Risk Factors for Early, Undesired Weaning"; C.-L. Dennis and K. McQueen, "The Relationship between Infant-Feeding Outcomes and Postpartum Depression: A Qualitative Systematic Review," *Pediatrics* 123, no. 4 (2009): e736-51, doi:10.1542/peds.2008–1629; S. Watkins et al., "Early Breastfeeding Experiences and Postpartum Depression," *Obstetrics and Gynecology* 118, no. 2 (2011): 214–21, doi:10.1097/AOG.0b013e3182260a2d; I. M. Paul et al., "Postpartum Anxiety and Maternal-Infant Health Outcomes," *Pediatrics* 131, no. 4 (2013): e1218-24, doi:10.1542/peds.2012–2147; E. M. Taveras et al., "Clinician Support and Psychosocial Risk Factors Associated with Breastfeeding Discontinuation," *Pediatrics* 112, no. 1 (2003): 108–15.
82. A. M. Stuebe, "Is Breastfeeding Promotion Bad for Mothers?" *Breastfeeding Medicine* (blog), February 21, 2011, http://bfmed.wordpress.com/2011/02/21/is-breastfeeding-promotion-bad-for-mothers.
83. M. Sapp and A. Vandeven, "Update on Childhood Sexual Abuse," *Current Opinion in Pediatrics* 17, no. 2 (2005): 258–64.
84. J. Coles, "Qualitative Study of Breastfeeding after Childhood Sexual Assault," *Journal of Human Lactation* 25, no. 3 (2009): 317–24, doi:10.1177/0890334409334926; C. T. Beck, "An Adult Survivor of Child Sexual Abuse and Her Breastfeeding Experience: A Case Study," *American Journal of Maternal/Child Nursing* 34, no. 2 (2009): 91–97, doi:10.1097/01.NMC.0000347302.85455.c8; S. K. Klingelhafer, "Sexual Abuse and Breastfeeding," *Journal of Human Lactation* 23, no. 2 (2007): 194–97, doi:10.1177/0890334407300387.
85. H. Stapleton, A. Fielder, and M. Kirkham, "Breast or Bottle? Eating Disordered Childbearing Women and Infant-Feeding Decisions," *Maternal and Child Nutrition* 4, no. 2 (2008): 106–20, doi:10.1111/j.1740–8709.2007.00121.x; E. Astrachan-Fletcher et al., "The Reciprocal Effects of Eating Disorders and the Postpartum Period: A Review of the Literature and Recommendations for Clinical Care," *Journal of Women's Health* 17, no. 2 (2008): 227–39, doi:10.1089/

jwh.2007.0550; E. Waugh and C. M. Bulik, "Offspring of Women with Eating Disorders," *International Journal of Eating Disorders* 25, no. 2 (1999): 123–33; S. Barston, "Of Human Bonding," in *Bottled Up: How the Way We Feed Babies Has Come to Define Motherhood, and Why It Shouldn't* (Berkeley, CA: University of California Press, 2012), 70–95.

86. Dennis and McQueen, "The Relationship between Infant-Feeding Outcomes and Postpartum Depression"; Mezzacappa and Katkin, "Breast-Feeding Is Associated with Reduced Perceived Stress."
87. E. Burns et al., "A Meta-Ethnographic Synthesis of Women's Experience of Breastfeeding," *Maternal and Child Nutrition* 6, no. 3 (2010): 201–19, doi:10.1111/j.1740-8709.2009.00209.x.
88. V. Schmied and D. Lupton, "Blurring the Boundaries: Breastfeeding and Maternal Subjectivity," *Sociology of Health and Illness* 23, no. 2 (2001): 234–50, doi:10.1111/1467-9566.00249.
89. A. Sheehan, V. Schmied, and L. Barclay, "Women's Experiences of Infant Feeding Support in the First 6 Weeks Post-Birth," *Maternal and Child Nutrition* 5, no. 2 (2009): 138–50, doi:10.1111/j.1740-8709.2008.00163.x; P. Hoddinott et al., "A Serial Qualitative Interview Study of Infant Feeding Experiences: Idealism Meets Realism," *BMJ Open* 2, no. 2 (2012): e000504, doi:10.1136/bmjopen-2011-000504.
90. Sheehan, Schmied, and Barclay, "Women's Experiences of Infant Feeding Support."
91. Michigan Department of Community Health, "WIC Helps Moms & Babies with Breastfeeding," accessed June 4, 2014, http://www.michigan.gov/mdch/0,4612,7-132-2942_4910_4919-13921—,00.html.
92. New York State Department of Health, "Breastfeed: Give the Gift of a Lifetime to Your Baby," accessed June 6, 2014, http://www.health.ny.gov/publications/3998.
93. D. Hegney, T. Fallon, and M. L. O'Brien, "Against All Odds: A Retrospective Case-Controlled Study of Women Who Experienced Extraordinary Breastfeeding Problems," *Journal of Clinical Nursing* 17, no. 9 (2008): 1182–92, doi:10.1111/j.1365-2702.2008.02300.x; Y. Hauck and V. Irurita, "Incompatible Expectations: The Dilemma of Breastfeeding Mothers," *Health Care for Women International* 24, no. 1 (2003): 62–78, doi:10.1080/07399330390170024.
94. E. Satter, "Understanding Your Newborn," in *Child of Mine: Feeding with Love and Good Sense* (Boulder, CO: Bull Publishing, 2000), 111–35.
95. J. P. Green, interview with the author.

CHAPTER 6. WHERE SHOULD YOUR BABY SLEEP?

1. Esmee McKee, email to the author, August 21, 2012.
2. G. A. Morelli et al., "Cultural Variation in Infants' Sleeping Arrangements: Questions of Independence," *Developmental Psychology* 28 (1992): 604–13.
3. S. Latz, A. W. Wolf, and B. Lozoff, "Cosleeping in Context: Sleep Practices

and Problems in Young Children in Japan and the United States," *Archives of Pediatrics and Adolescent Medicine* 153 (1999): 339–46.

4. B. Lozoff, G. L. Askew, and A. W. Wolf, "Cosleeping and Early Childhood Sleep Problems: Effects of Ethnicity and Socioeconomic Status," *Journal of Developmental Behavioral Pediatrics* 17 (1996): 9–15.
5. R. C. McCoy et al., "Frequency of Bed Sharing and Its Relationship to Breastfeeding," *Journal of Developmental and Behavioral Pediatrics* 25 (2004): 141–49; M. Willinger et al., "Trends in Infant Bed Sharing in the United States, 1993–2000: The National Infant Sleep Position Study," *Archives of Pediatrics and Adolescent Medicine* 157 (2003): 43–49; P. S. Blair, "The Prevalence and Characteristics Associated with Parent-Infant Bed-Sharing in England," *Archives of Disease in Childhood* 89, no. 12 (2004): 1106–10, doi:10.1136/adc.2003.038067.
6. McCoy et al., "Frequency of Bed Sharing"; Willinger et al., "Trends in Infant Bed Sharing."
7. S. L. Blunden, K. R. Thompson, and D. Dawson, "Behavioural Sleep Treatments and Night Time Crying in Infants: Challenging the Status Quo," *Sleep Medicine Reviews* 15 (2011): 327–34, doi:10.1016/j.smrv.2010.11.002; J. J. McKenna, H. L. Ball, and L. T. Gettler, "Mother-Infant Cosleeping, Breastfeeding and Sudden Infant Death Syndrome: What Biological Anthropology Has Discovered about Normal Infant Sleep and Pediatric Sleep Medicine," *American Journal of Physical Anthropology* 134, suppl. 45 (2007): 133–61, doi:10.1002/ajpa.20736; M. Small, *Our Babies, Ourselves: How Biology and Culture Shape the Way We Parent* (New York: Anchor Books, 1999).
8. American Academy of Pediatrics (AAP) Task Force on Sudden Infant Death Syndrome, "Policy Statement: SIDS and Other Sleep-Related Infant Deaths: Expansion of Recommendations for a Safe Infant Sleeping Environment," *Pediatrics* 128 (November 2011): 1030–39, doi:10.1542/peds.2011–2284.
9. Consumer Product Safety Commission (CPSC), "CPSC Warns against Placing Babies in Adult Beds; Study Finds 64 Deaths Each Year from Suffocation and Strangulation," 1999, http://www.cpsc.gov/CPSCPUB/PREREL/PRHTML99/99175.html.
10. Ask Dr. Sears, "Scientific Benefits of Co-Sleeping," accessed November 7, 2014, http://www.askdrsears.com/topics/sleep-problems/scientific-benefits-co-sleeping.
11. J. McKenna, "Frequently Asked Questions on Infant Sleep, SIDS Risks, Cosleeping, and Breastfeeding," Mother-Baby Behavioral Sleep Laboratory, accessed November 7, 2014, http://cosleeping.nd.edu/frequently-asked-questions/#40.
12. T. J. Mathews and M. F. MacDorman, *Infant Mortality Statistics from the 2009 Period Linked Birth/Infant Death Data Set*, National Health Statistics Reports, vol. 61, no. 8. (Hyattsville, MD: National Center for Health Statistics, 2013).
13. AAP Task Force on Sudden Infant Death Syndrome, "Technical Report: SIDS and Other Sleep-Related Infant Deaths: Expansion of Recommendations for a Safe Infant Sleeping Environment," *Pediatrics* 128 (November 2011): e1341-67, doi:10.1542/peds.2011–2285.

14. Centers for Disease Control and Prevention (CDC), "Infant Death Scene Investigation—SIDS and SUID," 2012, http://www.cdc.gov/sids/SceneInvestigation.htm.
15. P. G. Schnitzer, T. M. Covington, and H. K. Dykstra, "Sudden Unexpected Infant Deaths: Sleep Environment and Circumstances," *American Journal of Public Health* 102 (2012): 1204–12, doi:10.2105/AJPH.2011.300613.
16. Ibid.
17. F. R. Hauck et al., "Sleep Environment and the Risk of Sudden Infant Death Syndrome in an Urban Population: The Chicago Infant Mortality Study," *Pediatrics* 111, suppl. 1 (2003): 1207–14.
18. R. Y. Moon, R. S. C. Horne, and F. R. Hauck, "Sudden Infant Death Syndrome," *Lancet* 370 (2007): 1578–87.
19. H. C. Kinney et al., "The Brainstem and Serotonin in the Sudden Infant Death Syndrome," *Annual Review of Pathology: Mechanisms of Disease* 4, no. 1 (2009): 517–50, doi:10.1146/annurev.pathol.4.110807.092322; B. B. Randall et al., "Potential Asphyxia and Brainstem Abnormalities in Sudden and Unexpected Death in Infants," *Pediatrics* 132, no. 6 (2013): e1616-25, doi:10.1542/peds.2013–0700.
20. Kinney et al., "The Brainstem and Serotonin in the Sudden Infant Death Syndrome."
21. AAP Task Force on Sudden Infant Death Syndrome, "Technical Report."
22. D. E. Weese-Mayer et al., "Sudden Infant Death Syndrome: Review of Implicated Genetic Factors," *American Journal of Medical Genetics Part A* 143A, no. 8 (2007): 771–88, doi:10.1002/ajmg.a.31722.
23. F. L. Trachtenberg et al., "Risk Factor Changes for Sudden Infant Death Syndrome after Initiation of Back-to-Sleep Campaign," *Pediatrics* 129, no. 4 (2012): 630–38, doi:10.1542/peds.2011–1419.
24. R. Gilbert, "Infant Sleeping Position and the Sudden Infant Death Syndrome: Systematic Review of Observational Studies and Historical Review of Recommendations from 1940 to 2002," *International Journal of Epidemiology* 34, no. 4 (2005): 874–87, doi:10.1093/ije/dyi088.
25. Ibid.
26. A. Kahn et al., "Prone or Supine Body Position and Sleep Characteristics in Infants," *Pediatrics* 91, no. 6 (1993): 1112–15.
27. P. Franco et al., "Arousal from Sleep Mechanisms in Infants," *Sleep Medicine* 11, no. 7 (2010): 603–14, doi:10.1016/j.sleep.2009.12.014.
28. E. Pace, "Benjamin Spock, World's Pediatrician, Dies at 94," *New York Times*, March 17, 1998.
29. Gilbert, "Infant Sleeping Position."
30. R. G. Carpenter et al., "Sudden Unexplained Infant Death in 20 Regions in Europe: Case Control Study," *Lancet* 363, no. 9404 (2004): 185–91.
31. Ibid.
32. C. McGarvey et al., "An 8 Year Study of Risk Factors for SIDS: Bed-Sharing

versus Non-Bed-Sharing," *Archives of Disease in Childhood* 91, no. 4 (2006): 318–23, doi:10.1136/adc.2005.074674; P. J. Fleming et al., "Environment of Infants during Sleep and Risk of the Sudden Infant Death Syndrome: Results of 1993–5 Case-Control Study for Confidential Inquiry into Stillbirths and Deaths in Infancy," *BMJ* 313, no. 7051 (1996): 191–95, doi:10.1136/bmj.313.7051.191; R. Scragg et al., "Bed Sharing, Smoking, and Alcohol in the Sudden Infant Death Syndrome. New Zealand Cot Death Study Group," *BMJ* 307, no. 6915 (1993): 1312–18; M. M. Vennemann et al., "Sleep Environment Risk Factors for Sudden Infant Death Syndrome: The German Sudden Infant Death Syndrome Study," *Pediatrics* 123, no. 4 (2009): 1162–70, doi:10.1542/peds.2008–0505; M. H. Blabey and B. D. Gessner, "Infant Bed-Sharing Practices and Associated Risk Factors among Births and Infant Deaths in Alaska," *Public Health Reports* 124, no. 4 (2009): 527; Hauck et al., "Sleep Environment"; M. M. Vennemann et al., "Bed Sharing and the Risk of Sudden Infant Death Syndrome: Can We Resolve the Debate?" *Journal of Pediatrics* 160 (2012): 44–48, doi:10.1016/j.jpeds.2011.06.052.

33. P. S. Blair et al., "Babies Sleeping with Parents: Case-Control Study of Factors Influencing the Risk of the Sudden Infant Death Syndrome," *BMJ* 319, no. 7223 (1999): 1457–62; P. S. Blair et al., "Hazardous Cosleeping Environments and Risk Factors Amenable to Change: Case-Control Study of SIDS in South West England," *BMJ* 339 (2009): b3666, doi:10.1136/bmj.b3666.
34. Blair et al., "Babies Sleeping with Parents"; Carpenter et al., "Sudden Unexplained Infant Death"; McGarvey et al., "An 8 Year Study of Risk Factors"; D. Tappin, R. Ecob, and H. Brooke, "Bedsharing, Roomsharing, and Sudden Infant Death Syndrome in Scotland: A Case-Control Study," *Journal of Pediatrics* 147, no. 1 (2005): 32–37, doi:10.1016/j.jpeds.2005.01.035.
35. Blair et al., "Babies Sleeping with Parents"; Blair et al., "Hazardous Cosleeping Environments"; P. S. Blair, "Sudden Infant Death Syndrome Epidemiology and Bed Sharing," *Pediatric Child Health* 11, suppl. A (2006): 29-31A.
36. R. Carpenter et al., "Bed Sharing When Parents Do Not Smoke: Is There a Risk of SIDS? An Individual Level Analysis of Five Major Case-Control Studies," *BMJ Open* 3, no. 5 (2013), doi:10.1136/bmjopen-2012–002299.
37. UNICEF UK Baby Friendly Initiative, "Baby Friendly Initiative Statement on New Bed Sharing Research," May 21, 2013, http://www.unicef.org.uk/BabyFriendly/News-and-Research/News/UNICEF-UK-Baby-Friendly-Initiative-statement-on-new-bed-sharing-research; Infant Sleep Information Source, "Response to New Analysis of Bed-Sharing and SIDS: Carpenter et al. (2013) in BMJ Open," *ISIS Online*, May 20, 2013, http://www.isisonline.org.uk/news/?itemno=17810.
38. R. S. Horne et al., "Peer Review History for: Bed Sharing When Parents Do Not Smoke: Is There a Risk of SIDS? An Individual Level Analysis of Five Major Case-Control Studies by Carpenter et al.," *BMJ Open*, 2013, http://bmjopen.bmj.com/content/3/5/e002299.
39. P. S. Blair et al., "Bed-Sharing in the Absence of Hazardous Circumstances:

Is There a Risk of Sudden Infant Death Syndrome? An Analysis from Two Case-Control Studies Conducted in the UK," *PLoS ONE* 9, no. 9 (September 19, 2014): e107799, doi:10.1371/journal.pone.0107799.

40. S. A. Baddock et al., "Hypoxic and Hypercapnic Events in Young Infants during Bed-Sharing," *Pediatrics*, July 16, 2012, doi:10.1542/peds.2011–3390; S. A. Baddock, "Differences in Infant and Parent Behaviors during Routine Bed Sharing Compared with Cot Sleeping in the Home Setting," *Pediatrics* 117, no. 5 (May 1, 2006): 1599–1607, doi:10.1542/peds.2005–1636; S. A. Baddock et al., "Sleep Arrangements and Behavior of Bed-Sharing Families in the Home Setting," *Pediatrics* 119 (January 2007): e200-207, doi:10.1542/peds.2006–0744.
41. H. Ball, "Airway Covering during Bed-Sharing," *Child: Care, Health, and Development* 35, no. 5 (September 2009): 728–37, doi:10.1111/j.1365–2214.2009.00979.x.
42. D. P. Davies, "Cot Death in Hong Kong: A Rare Problem?" *Lancet* 2, no. 8468 (1985): 1346–49.
43. N. N. Lee et al., "Sudden Infant Death Syndrome in Hong Kong: Confirmation of Low Incidence," BMJ 298, no. 6675 (1989): 721.
44. Davies, "Cot Death in Hong Kong."
45. E. A. Nelson et al., "International Child Care Practices Study: Infant Sleeping Environment," *Early Human Development* 62, no. 1 (2001): 43–55.
46. P. G. Tuohy, A. M. Counsell, and D. C. Geddis, "Sociodemographic Factors Associated with Sleeping Position and Location," *Archives of Disease in Childhood* 69, no. 6 (1993): 664–66; S. Farooqi, I. J. Perry, and D. G. Beevers, "Ethnic Differences in Infant-Rearing Practices and Their Possible Relationship to the Incidence of Sudden Infant Death Syndrome (SIDS)," *Paediatric and Perinatal Epidemiology* 7, no. 3 (1993): 245–52; M. Gantley, D. P. Davies, and A. Murcott, "Sudden Infant Death Syndrome: Links with Infant Care Practices," BMJ 306, no. 6869 (1993): 16–20.
47. Rachel Y. Moon, telephone interview with the author, April 15, 2013.
48. Carpenter et al., "Sudden Unexplained Infant Death."
49. Tappin, Ecob, and Brooke, "Bedsharing, Roomsharing"; Blair et al., "Hazardous Cosleeping Environments."
50. P. Blair and S. Inch, "The Health Professional's Guide to: 'Caring for Your Baby at Night,'" UNICEF U.K. Baby Friendly Initiative, 2011, https://www.unicef.org.uk/Documents/Baby_Friendly/Leaflets/HPs_Guide_to_Coping_At_Night_Final.pdf.
51. M. Small, "A Reasonable Sleep," in Small, *Our Babies, Ourselves*, 109–37.
52. S. S. Mosko, C. A. Richard, and J. J. McKenna, "Infant Arousals during Mother-Infant Bed Sharing: Implications for Infant Sleep and Sudden Infant Death Syndrome Research," *Pediatrics* 100, no. 5 (1997): 841–49, doi:10.1542/peds.100.5.841; S. S. Mosko et al., "Infant Sleep Architecture during Bedsharing and Possible Implications for SIDS," *Sleep* 19, no. 9 (1996): 677–84.
53. S. S. Mosko, C. A. Richard, and J. J. McKenna, "Maternal Sleep and Arousals during Bedsharing with Infants," *Sleep* 20 (1997): 142–50.

54. J. J. McKenna, S. S. Mosko, and C. A. Richard, "Bedsharing Promotes Breastfeeding," *Pediatrics* 100 (1997): 214–19.
55. C. A. Richard, S. S. Mosko, and J. J. McKenna, "Apnea and Periodic Breathing in Bed-Sharing and Solitary Sleeping Infants," *Journal of Applied Physiology* 84, no. 4 (1998): 1374–80.
56. J. J. McKenna and T. McDade, "Why Babies Should Never Sleep Alone: A Review of the Co-Sleeping Controversy in Relation to SIDS, Bedsharing and Breast Feeding," *Paediatric Respiratory Reviews* 6 (2005): 134–52, doi:10.1016/j.prrv.2005.03.006.
57. Blair et al., "Bed-Sharing in the Absence of Hazardous Circumstances."
58. K. Hinde and L. A Milligan, "Primate Milk: Proximate Mechanisms and Ultimate Perspectives," *Evolutionary Anthropology: Issues, News, and Reviews* 20, no. 1 (2011): 9–23, doi:10.1002/evan.20289.
59. H. L. Ball, "Breastfeeding, Bed-Sharing, and Infant Sleep," *Birth* 30, no. 3 (2003): 181–88; Y. Huang et al., "Influence of Bedsharing Activity on Breastfeeding Duration among US Mothers," *JAMA Pediatrics* 167, no. 11 (2013): 1038–44, doi:10.1001/jamapediatrics.2013.2632.
60. P. S. Blair, J. Heron, and P. J. Fleming, "Relationship between Bed Sharing and Breastfeeding: Longitudinal, Population-Based Analysis," *Pediatrics* 126, no. 5 (2010): e1119-26, doi:10.1542/peds.2010–1277.
61. Blair et al., "Hazardous Cosleeping Environments."
62. K. Kendall-Tackett, Z. Cong, and T. W. Hale, "Mother-Infant Sleep Locations and Nighttime Feeding Behavior," *Clinical Lactation* 1, no. 1 (2010): 27–30.
63. McGarvey et al., "An 8 Year Study of Risk Factors"; Fleming et al., "Environment of Infants during Sleep."
64. AAP Task Force on Sudden Infant Death Syndrome, "Technical Report."
65. P. Franco et al., "The Influence of a Pacifier on Infants' Arousals from Sleep," *Journal of Pediatrics* 136, no. 6 (2000): 775–79, doi:10.1067/mpd.2000.105365; R. S. C. Horne et al., "Comparison of Evoked Arousability in Breast and Formula Fed Infants," *Archives of Disease in Childhood* 89, no. 1 (2004): 22–25.
66. F. R. Hauck et al., "Breastfeeding and Reduced Risk of Sudden Infant Death Syndrome: A Meta-analysis," *Pediatrics* 128, no. 1 (2011): 103–10, doi:10.1542/peds.2010–3000.
67. AAP Task Force on Sudden Infant Death Syndrome, "Policy Statement."
68. Moon, interview with the author.

CHAPTER 7. IN SEARCH OF A GOOD NIGHT'S SLEEP

1. I. Iglowstein et al., "Sleep Duration from Infancy to Adolescence: Reference Values and Generational Trends," *Pediatrics* 111, no. 2 (2003): 302–7, doi:10.1542/peds.111.2.302.
2. B. Cavell, "Gastric Emptying in Infants Fed Human Milk or Infant Formula," *Acta Paediatrica Scandinavica* 70 (1981): 639–41; E. Sievers et al., "Feeding Patterns in Breast-Fed and Formula-Fed Infants," *Annals of Nutrition and Metabolism* 46, no. 6 (2002): 243–48, doi:10.1159/000066498.

3. M. De Carvalho, S. Robertson, and A. Friedman, "Effect of Frequent Breast-Feeding on Early Milk Production and Infant Weight Gain," *Pediatrics* 72, no. 3 (1983): 307–11.
4. A. W. de Weerd and R. A. van den Bossche, "The Development of Sleep during the First Months of Life," *Sleep Medicine Reviews* 7, no. 2 (2003): 179–91, doi:10.1053/smrv.2002.0198.
5. Ibid.; T. F. Anders, "Organization and Development of Sleep in Early Life," in *Encyclopedia on Early Childhood Development*, ed. R. E. Tremblay, M. Boivin, and R. Peters (Montreal: Centre of Excellence for Early Childhood Development and Strategic Knowledge Cluster on Early Child Development, 2010), 1–8, http://child-encyclopedia.com/pages/PDF/sleeping_behaviour.pdf#page=23.
6. S. Coons and C. Guilleminault, "Development of Sleep-Wake Patterns and Non-Rapid Eye Movement Sleep Stages during the First Six Months of Life in Normal Infants," *Pediatrics* 69, no. 6 (1982): 793–98; de Weerd and van den Bossche, "The Development of Sleep"; Anders, "Organization and Development of Sleep."
7. Sievers et al., "Feeding Patterns."
8. R. Y. Moore, "Suprachiasmatic Nucleus in Sleep-Wake Regulation," *Sleep Medicine* 8 (2007): 27–33, doi:10.1016/j.sleep.2007.10.003.
9. K. McGraw et al., "The Development of Circadian Rhythms in a Human Infant," *Sleep* 22, no. 3 (1999): 303–10.
10. Ibid.
11. J. Ardura et al., "Emergence and Evolution of the Circadian Rhythm of Melatonin in Children," *Hormone Research* 59, no. 2 (2003): 66–72, doi:10.1159/000068571; M. Mirmiran, Y. G. Maas, and R. L. Ariagno, "Development of Fetal and Neonatal Sleep and Circadian Rhythms," *Sleep Medicine Reviews* 7, no. 4 (2003): 321–34, doi:10.1053/smrv.2002.0243.
12. R. J. Custodio et al., "The Emergence of the Cortisol Circadian Rhythm in Monozygotic and Dizygotic Twin Infants: The Twin-Pair Synchrony," *Clinical Endocrinology* 66, no. 2 (2007): 192–97, doi:10.1111/j.1365–2265.2006.02706.x.
13. A. Cohen Engler et al., "Breastfeeding May Improve Nocturnal Sleep and Reduce Infantile Colic: Potential Role of Breast Milk Melatonin," *European Journal of Pediatrics* 171, no. 4 (2011): 729–32, doi:10.1007/s00431-011–1659–3; J. Cubero et al., "The Circadian Rhythm of Tryptophan in Breast Milk Affects the Rhythms of 6-Sulfatoxymelatonin and Sleep in Newborn," *Neuroendocrinology Letters* 26, no. 6 (2005): 657–61; H. Illnerová, M. Buresová, and J. Presl, "Melatonin Rhythm in Human Milk," *Journal of Clinical Endocrinology and Metabolism* 77, no. 3 (1993): 838–41.
14. Y. Harrison, "The Relationship between Daytime Exposure to Light and Night-Time Sleep in 6–12-Week-Old Infants," *Journal of Sleep Research* 13 (2004): 345–52, doi:10.1111/j.1365–2869.2004.00435.x.
15. I. C. McMillen et al., "Development of Circadian Sleep-Wake Rhythms in Preterm and Full-Term Infants," *Pediatric Research* 29, no. 4 (1991): 381–84.

16. McGraw et al., "The Development of Circadian Rhythms."
17. Coons and Guilleminault, "Development of Sleep-Wake Patterns."
18. J. A. Mindell et al., "A Nightly Bedtime Routine: Impact on Sleep in Young Children and Maternal Mood," *Sleep* 32 (2009): 599–606.
19. H. L. Ball, "Breastfeeding, Bed-Sharing, and Infant Sleep," *Birth* 30, no. 3 (2003): 181–88; J. M. T. Henderson et al., "Sleeping through the Night: The Consolidation of Self-Regulated Sleep across the First Year of Life," *Pediatrics* 126, no. 5 (2010): e1081-87, doi:10.1542/peds.2010–0976.
20. E. Touchette et al., "Genetic and Environmental Influences on Daytime and Nighttime Sleep Duration in Early Childhood," *Pediatrics* 131, no. 6 (2013): e1874-80, doi:10.1542/peds.2012–2284.
21. J. Eaton-Evans and A. E. Dugdale, "Sleep Patterns of Infants in the First Year of Life," *Archives of Disease in Childhood* 63, no. 6 (1988): 647–49; T. F. Anders and M. Keener, "Developmental Course of Nighttime Sleep-Wake Patterns in Full-Term and Premature Infants during the First Year of Life. I," *Sleep* 8, no. 3 (1985): 173–92; C. W. DeLeon and K. H. Karraker, "Intrinsic and Extrinsic Factors Associated with Night Waking in 9-Month-Old Infants," *Infant Behavior and Development* 30 (December 2007): 596–605, doi:10.1016/j.infbeh.2007.03.009; A. Scher, R. Epstein, and E. Tirosh, "Stability and Changes in Sleep Regulation: A Longitudinal Study from 3 Months to 3 Years," *International Journal of Behavioral Development* 28, no. 3 (2004): 268–74, doi:10.1080/01650250344000505.
22. K. So, T. M. Adamson, and R. S. Horne, "The Use of Actigraphy for Assessment of the Development of Sleep/Wake Patterns in Infants during the First 12 Months of Life," *Journal of Sleep Research* 16, no. 2 (2007): 181–87; M. M. Burnham et al., "Nighttime Sleep-Wake Patterns and Self-soothing from Birth to One Year of Age: A Longitudinal Intervention Study," *Journal of Child Psychology and Psychiatry* 43 (2002): 713–25.
23. National Sleep Foundation, *Sleep in America Poll: Children and Sleep*, 2004, http://www.sleepfoundation.org/article/sleep-america-polls/2004-children-and-sleep.
24. Ball, "Breastfeeding, Bed-Sharing, and Infant Sleep."
25. J. J. McKenna and L. E. Volpe, "Sleeping with Baby: An Internet-Based Sampling of Parental Experiences, Choices, Perceptions, and Interpretations in a Western Industrialized Context," *Infant and Child Development* 16, no. 4 (2007): 359–85, doi:10.1002/icd.525.
26. M. F. Elias et al., "Sleep/Wake Patterns of Breast-Fed Infants in the First 2 Years of Life," *Pediatrics* 77 (March 1986): 322–29; F. R. Hauck et al., "Infant Sleeping Arrangements and Practices during the First Year of Life," *Pediatrics* 122, suppl. (2008): S113-120, doi:10.1542/peds.2008-13150; A. Mao et al., "A Comparison of the Sleep-Wake Patterns of Cosleeping and Solitary-Sleeping Infants," *Child Psychiatry and Human Development* 35 (2004): 95–105, doi:10.1007/s10578-004–1879–0; J. A. Mindell et al., "Cross-Cultural Differences in Infant and Tod-

dler Sleep," *Sleep Medicine* 11 (2010): 274–80, doi:10.1016/j.sleep.2009.04.012; National Sleep Foundation, *Sleep in America Poll.*

27. Mao et al., "A Comparison of the Sleep-Wake Patterns."
28. Mindell et al., "Cross-Cultural Differences in Infant and Toddler Sleep."
29. Hauck et al., "Infant Sleeping Arrangements"; McKenna and Volpe, "Sleeping with Baby."
30. S. Mosko, C. Richard, and J. McKenna, "Maternal Sleep and Arousals during Bedsharing with Infants," *Sleep* 20 (1997): 142–50.
31. Elias et al., "Sleep/Wake Patterns of Breast-Fed Infants."
32. Hauck et al., "Infant Sleeping Arrangements."
33. Elias et al., "Sleep/Wake Patterns of Breast-Fed Infants."
34. Ibid.
35. T. Pinilla and L. L. Birch, "Help Me Make It through the Night: Behavioral Entrainment of Breast-Fed Infants' Sleep Patterns," *Pediatrics* 91 (February 1993): 436–44.
36. Douglas M. Teti, telephone interview with the author, February 11, 2014; D. M. Teti et al., "Maternal Emotional Availability at Bedtime Predicts Infant Sleep Quality," *Journal of Family Psychology* 24 (2010): 307–15, doi:10.1037/a0019306.
37. Leita Dzubay, telephone interview with the author, August 23, 2012.
38. Thomas F. Anders, telephone interview with the author, April 26, 2013.
39. Ibid.
40. T. F. Anders, "Home-Recorded Sleep in 2- and 9-Month-Old Infants," *Journal of the American Academy of Child Psychiatry* 17, no. 3 (1978): 421–32.
41. T. F. Anders, "Night-Waking in Infants during the First Year of Life," *Pediatrics* 63 (1979): 860–64.
42. Anders and Keener, "Developmental Course of Nighttime Sleep-Wake Patterns."
43. Anders, interview with the author.
44. T. F. Anders, L. F. Halpern, and J. Hua, "Sleeping through the Night: A Developmental Perspective," *Pediatrics* 90 (1992): 554–60.
45. Ibid.; W. Anuntaseree et al., "Night Waking in Thai Infants at 3 Months of Age: Association between Parental Practices and Infant Sleep," *Sleep Medicine* 9, no. 5 (2008): 564–71, doi:10.1016/j.sleep.2007.07.009; Burnham et al., "Nighttime Sleep-Wake Patterns"; É. Touchette et al., "Factors Associated with Fragmented Sleep at Night across Early Childhood," *Archives of Pediatrics and Adolescent Medicine* 159, no. 3 (2005): 242; J. A. Mindell et al., "Parental Behaviors and Sleep Outcomes in Infants and Toddlers: A Cross-Cultural Comparison," *Sleep Medicine* 11 (2010): 393–99, doi:10.1016/j.sleep.2009.11.011.
46. Anders, interview with the author.
47. Burnham et al., "Nighttime Sleep-Wake Patterns."
48. M. A. Keener, C. H. Zeanah, and T. F. Anders, "Infant Temperament, Sleep

Organization, and Nighttime Parental Interventions," *Pediatrics* 81 (1988): 762–71.

49. Pinilla and Birch, "Help Me Make It through the Night"; A. Wolfson, P. Lacks, and A. Futterman, "Effects of Parent Training on Infant Sleeping Patterns, Parents' Stress, and Perceived Parental Competence," *Journal of Consulting and Clinical Psychology* 60 (1992): 41–48.
50. Anders, Halpern, and Hua, "Sleeping through the Night"; B. L. Goodlin-Jones et al., "Night Waking, Sleep-Wake Organization, and Self-Soothing in the First Year of Life," *Journal of Developmental and Behavioral Pediatrics* 22 (2001): 226–33.
51. Burnham et al., "Nighttime Sleep-Wake Patterns."
52. D. W. Winnicott, "Transitional Objects and Transitional Phenomena: A Study of the First Not-Me Possession," *International Journal of Psycho-Analysis* 34 (1953): 89–97.
53. Anders, interview with the author.
54. G. A. Morelli et al., "Cultural Variation in Infants' Sleeping Arrangements: Questions of Independence," *Developmental Psychology* 28 (1992): 604–13; K. M. Hong and B. D. Townes, "Infants' Attachment to Inanimate Objects," *Journal of the American Academy of Child and Adolescent Psychiatry* 15 (1976): 49–61.
55. A. W. Wolf and B. Lozoff, "Object Attachment, Thumbsucking, and the Passage to Sleep," *Journal of the American Academy of Child and Adolescent Psychiatry* 28, no. 2 (1989): 287–92.
56. W. Sears and M. Sears, *The Baby Book: Everything You Need to Know about Your Baby from Birth to Age Two*, rev. and updated ed. (New York: Little, Brown, and Co., 2003).
57. M. B. Ramamurthy et al., "Effect of Current Breastfeeding on Sleep Patterns in Infants from Asia-Pacific Region," *Journal of Paediatrics and Child Health* 48, no. 8 (2012): 669–74, doi:10.1111/j.1440–1754.2012.02453.x; Keener, Zeanah, and Anders, "Infant Temperament."
58. S. Latz, A. W. Wolf, and B. Lozoff, "Cosleeping in Context: Sleep Practices and Problems in Young Children in Japan and the United States," *Archives of Pediatrics and Adolescent Medicine* 153 (1999): 339–46.
59. L. Tikotzky and A. Sadeh, "Maternal Sleep-Related Cognitions and Infant Sleep: A Longitudinal Study from Pregnancy through the 1st Year," *Child Development* 80, no. 3 (2009): 860–74.
60. Ibid.
61. L. Tikotzky and L. Shaashua, "Infant Sleep and Early Parental Sleep-Related Cognitions Predict Sleep in Pre-School Children," *Sleep Medicine* 13, no. 2 (2012): 185–92, doi:10.1016/j.sleep.2011.07.013.
62. Tikotzky and Sadeh, "Maternal Sleep-Related Cognitions."
63. Anders, Halpern, and Hua, "Sleeping through the Night."
64. Touchette et al., "Genetic and Environmental Influences."

65. D. Fields and A. Brown, *Baby 411: Clear Answers and Smart Advice for Your Baby's First Year*, 3rd ed. (Boulder, CO: Windsor Peak Press, 2008).
66. H. Hiscock et al., "Improving Infant Sleep and Maternal Mental Health: A Cluster Randomised Trial," *Archives of Disease in Childhood* 92 (2007): 952–58, doi:10.1136/adc.2006.099812; H. Hiscock and M. Wake, "Randomised Controlled Trial of Behavioural Infant Sleep Intervention to Improve Infant Sleep and Maternal Mood," *BMJ* 324 (2002): 1062–65; J. A. Mindell et al., "Behavioral Treatment of Bedtime Problems and Night Wakings in Infants and Young Children," *Sleep* 29 (2006): 1263–76.
67. Mindell et al., "Behavioral Treatment of Bedtime Problems."
68. Jodi A. Mindell, email to the author, May 6, 2013.
69. D. Narvaez, "Moral Landscapes: Dangers of 'Crying It Out,'" *Psychology Today*, December 11, 2011, http://www.psychologytoday.com/blog/moral-landscapes/201112/dangers-crying-it-out.
70. J. P. Shonkoff and A. S. Garner, "The Lifelong Effects of Early Childhood Adversity and Toxic Stress," *Pediatrics* 129 (2012): e232-46, doi:10.1542/peds.2011–2663.
71. J. Martin et al., "Adverse Associations of Infant and Child Sleep Problems and Parent Health: An Australian Population Study," *Pediatrics* 119 (2007): 947–55, doi:10.1542/peds.2006–2569; Hiscock et al., "Improving Infant Sleep."
72. A. M. Medina, C. L. Lederhos, and T. A. Lillis, "Sleep Disruption and Decline in Marital Satisfaction across the Transition to Parenthood," *Families, Systems, and Health* 27, no. 2 (2009): 153–60, doi:10.1037/a0015762.
73. A. M. Williamson and A. M. Feyer, "Moderate Sleep Deprivation Produces Impairments in Cognitive and Motor Performance Equivalent to Legally Prescribed Levels of Alcohol Intoxication," *Occupational and Environmental Medicine* 57, no. 10 (2000): 649–55.
74. W. Middlemiss et al., "Asynchrony of Mother-Infant Hypothalamic-Pituitary-Adrenal Axis Activity Following Extinction of Infant Crying Responses Induced during the Transition to Sleep," *Early Human Development* 88, no. 4 (2011): 227–32.
75. Anders, interview with the author.

CHAPTER 8. VACCINES AND YOUR CHILD

1. Margaret Jordan Green, interview with the author, Gainesville, FL, March 1, 2014.
2. W. J. Moss and D. E. Griffin, "Measles," *Lancet* 379, no. 9811 (2012): 153–64, doi:10.1016/S0140-6736(10)62352–5.
3. Ibid.
4. R. Buchanan and D. J. Bonthius, "Measles Virus and Associated Central Nervous System Sequelae," *Seminars in Pediatric Neurology* 19, no. 3 (2012): 107–14, doi:10.1016/j.spen.2012.02.003.
5. Centers for Disease Control and Prevention (CDC), "MMWR Summary of

Notifiable Diseases, United States, 1993," October 21, 1994, http://www.cdc.gov/mmwr/preview/mmwrhtml/00035381.htm#00000865.htm.

6. A. Allen, "Battling Measles, Remodeling Society," in *Vaccine: The Controversial Story of Medicine's Greatest Lifesaver* (New York: W. W. Norton, 2007), 215–47.
7. National Communicable Disease Center (U.S.), "Reported Incidence of Notifiable Diseases in the United States, 1966," 1967, http://stacks.cdc.gov/view/cdc/615.
8. C. E. Koop, *Measles Continues to Spread in the U.S.: 1916* (video), Measles Timeline, Vaccine Makers Project, accessed February 5, 2014, http://www.historyofvaccines.org/content/timelines/measles.
9. J. P. Baker, "The First Measles Vaccine," *Pediatrics* 128, no. 3 (2011): 435–37, doi:10.1542/peds.2011–1430.
10. Nobel Media, "John F. Enders—Biographical," Nobelprize.org, 2013, http://www.nobelprize.org/nobel_prizes/medicine/laureates/1954/enders-bio.html.
11. S. L. Katz, *Thomas Peebles Isolates the Measles Virus: 1954* (video), Measles Timeline, Vaccine Makers Project, accessed June 17, 2014, http://www.historyofvaccines.org/content/timelines/measles.
12. Baker, "The First Measles Vaccine."
13. Ibid.; S. L. Katz, "The History of Measles Vaccine and Attempts to Control Measles," in *Microbe Hunters: Then and Now* (Bloomington, IL: Medi-Ed Press, 1996), 69–76.
14. P. A. Offit, "The Destroying Angel," in *Vaccinated: One Man's Quest to Defeat the World's Deadliest Diseases* (New York: Harper Collins, 2007), 44–56.
15. Allen, "Battling Measles, Remodeling Society."
16. W. Atkinson, S. Wolfe, and J. Hamborsky, eds., *Epidemiology and Prevention of Vaccine-Preventable Diseases*, 12th ed. (Washington, DC: Public Health Foundation, 2012), http://www.cdc.gov/vaccines/pubs/pinkbook/meas.html#secular.
17. A. P. Fiebelkorn et al., "Measles in the United States during the Postelimination Era," *Journal of Infectious Diseases* 202, no. 10 (November 15, 2010): 1520–28, doi:10.1086/656914.
18. CDC, "Measles: Cases and Outbreaks," 2014, http://www.cdc.gov/measles/cases-outbreaks.html.
19. World Health Organization (WHO), *Global Control and Regional Elimination of Measles, 2000–2012*, Weekly Epidemiological Record (Geneva: WHO, February 7, 2014).
20. CDC, "Child and Adolescent Immunization Schedules," 2014, http://www.cdc.gov/vaccines/schedules/hcp/imz/child-adolescent.html.
21. S. W. Roush, T. V. Murphy, and Vaccine-Preventable Disease Table Working Group, "Historical Comparisons of Morbidity and Mortality for Vaccine-Preventable Diseases in the United States," *JAMA* 298, no. 18 (2007): 2155–63, doi:10.1001/jama.298.18.2155.
22. F. Zhou et al., "Economic Evaluation of the Routine Childhood Immuniza-

tion Program in the United States, 2009," *Pediatrics* 133, no. 4 (2014): 577–85, doi:10.1542/peds.2013-0698.

23. J. F. Seward et al., "Varicella Disease after Introduction of Varicella Vaccine in the United States, 1995–2000," *JAMA* 287, no. 5 (2002): 606–11, doi:10.1001/jama.287.5.606; D. Guris et al., "Changing Varicella Epidemiology in Active Surveillance Sites—United States, 1995–2005," *Journal of Infectious Diseases* 197, suppl. 2 (2008): S71-75, doi:10.1086/522156.
24. R. I. Glass et al., "Rotavirus Vaccines: Successes and Challenges," *Journal of Infection* 68, suppl. 1 (2014): S9-18, doi:10.1016/j.jinf.2013.09.010.
25. C. A. Panozzo et al., "Direct, Indirect, Total, and Overall Effectiveness of the Rotavirus Vaccines for the Prevention of Gastroenteritis Hospitalizations in Privately Insured US Children, 2007–2010," *American Journal of Epidemiology* 179, no. 7 (2014): 895–909, doi:10.1093/aje/kwu001.
26. D. R. Feikin et al., "Individual and Community Risks of Measles and Pertussis Associated with Personal Exemptions to Immunization," *JAMA* 284, no. 24 (2000): 3145–50, doi:10.1001/jama.284.24.3145; J. M. Glanz et al., "Parental Refusal of Pertussis Vaccination Is Associated with an Increased Risk of Pertussis Infection in Children," *Pediatrics* 123, no. 6 (2009): 1446–51, doi:10.1542/peds.2008-2150.
27. J. E. Atwell et al., "Nonmedical Vaccine Exemptions and Pertussis in California, 2010," *Pediatrics* 132, no. 4 (2013): 624–30, doi:10.1542/peds.2013-0878; S. B. Omer et al., "Nonmedical Exemptions to School Immunization Requirements: Secular Trends and Association of State Policies with Pertussis Incidence," *JAMA* 296, no. 14 (2006): 1757–63, doi:10.1001/jama.296.14.1757; S. B. Omer et al., "Geographic Clustering of Nonmedical Exemptions to School Immunization Requirements and Associations with Geographic Clustering of Pertussis," *American Journal of Epidemiology* 168, no. 12 (2008): 1389–96, doi:10.1093/aje/kwn263; J. M. Glanz et al., "Parental Refusal of Varicella Vaccination and the Associated Risk of Varicella Infection in Children," *Archives of Pediatrics and Adolescent Medicine* 164, no. 1 (2010): 66–70, doi:10.1001/archpediatrics.2009.244; J. M. Glanz et al., "Parental Decline of Pneumococcal Vaccination and Risk of Pneumococcal Related Disease in Children," *Vaccine* 29, no. 5 (2011): 994–99, doi:10.1016/j.vaccine.2010.11.085.
28. R. S. Barlow et al., "Vaccinated Children and Adolescents with Pertussis Infections Have Decreased Illness Severity and Duration, Oregon 2010–2012," *Clinical Infectious Diseases* 58, no. 11 (2014): 1523–29, doi:10.1093/cid/ciu156; P. Mitchell et al., "Previous Vaccination Modifies Both the Clinical Disease and Immunological Features in Children with Measles," *Journal of Primary Health Care* 5, no. 2 (2013): 93–98.
29. P. Fine, K. Eames, and D. L. Heymann, "'Herd Immunity': A Rough Guide," *Clinical Infectious Diseases* 52, no. 7 (2011): 911–16, doi:10.1093/cid/cir007; H. Rashid, G. Khandaker, and R. Booy, "Vaccination and Herd Immunity: What

More Do We Know?" *Current Opinion in Infectious Diseases* 25, no. 3 (2012): 243–49, doi:10.1097/QCO.0b013e328352f727.

30. P. G. Smith, "Concepts of Herd Protection and Immunity," *Procedia in Vaccinology* 2, no. 2 (2010): 134–39, doi:10.1016/j.provac.2010.07.005.
31. D. E. Sugerman et al., "Measles Outbreak in a Highly Vaccinated Population, San Diego, 2008: Role of the Intentionally Undervaccinated," *Pediatrics* 125, no. 4 (2010): 747–55.
32. Ibid.
33. Ibid.
34. W. A. Orenstein, "The Role of Measles Elimination in Development of a National Immunization Program," *Pediatric Infectious Disease Journal* 25, no. 12 (2006): 1093–1101, doi:10.1097/01.inf.0000246840.13477.28.
35. S. S. Chaves et al., "Varicella in Infants after Implementation of the US Varicella Vaccination Program," *Pediatrics* 128, no. 6 (2011): 1071–77, doi:10.1542/peds.2011-0017.
36. M. M. Patel et al., "Fulfilling the Promise of Rotavirus Vaccines: How Far Have We Come since Licensure?" *Lancet Infectious Diseases* 12, no. 7 (2012): 561–70, doi:10.1016/S1473-3099(12)70029-4; P. A. Gastañaduy et al., "Gastroenteritis Hospitalizations in Older Children and Adults in the United States before and after Implementation of Infant Rotavirus Vaccination," *JAMA* 310, no. 8 (2013): 851–53, doi:10.1001/jama.2013.170800.
37. Rashid, Khandaker, and Booy, "Vaccination and Herd Immunity."
38. Feikin et al., "Individual and Community Risks."
39. CDC, "Prevention of Measles, Rubella, Congenital Rubella Syndrome, and Mumps, 2013: Summary Recommendations of the Advisory Committee on Immunization Practices (ACIP)," *MMWR Recommendations and Reports* 62, no. 4 (2013), http://www.cdc.gov/mmwr/preview/mmwrhtml/rr6204a1.htm; V. Demicheli et al., "Vaccines for Measles, Mumps and Rubella in Children," *Cochrane Database of Systematic Reviews* 2 (2012): CD004407, doi:10.1002/14651858.CD004407.pub3.
40. N. Principi and S. Esposito, "Vaccines and Febrile Seizures," *Expert Review of Vaccines* 12, no. 8 (2013): 885–92, doi:10.1586/14760584.2013.814781.
41. Atkinson, Wolfe, and Hamborsky, *Epidemiology and Prevention of Vaccine-Preventable Diseases.*
42. K. Bohlke et al., "Risk of Anaphylaxis after Vaccination of Children and Adolescents," *Pediatrics* 112, no. 4 (2003): 815–20, doi:10.1542/peds.112.4.815.
43. Institute of Medicine (IOM), "Reports," 2014, http://www.iom.edu/Reports.aspx.
44. IOM, *Childhood Immunization Schedule and Safety: Stakeholder Concerns, Scientific Evidence, and Future Studies* (Washington, DC: National Academies Press, 2013), http://www.iom.edu/Reports/2013/The-Childhood-Immunization-Schedule-and-Safety.aspx.
45. J. L. Schwartz, "The First Rotavirus Vaccine and the Politics of Acceptable

Risk," *Milbank Quarterly* 90, no. 2 (2012): 278–310, doi:10.1111/j.1468-0009.2012.00664.x.

46. U.S. Food and Drug Administration (FDA), "Vaccine Product Approval Process," U.S. Food and Drug Administration, Vaccines, Blood, and Biologics, 2009, http://www.fda.gov/biologicsbloodvaccines/developmentapprovalprocess/biologicslicenseapplicationsblaprocess/ucm133096.htm; Schwartz, "The First Rotavirus Vaccine."
47. J. C. Smith, D. E. Snider, and L. K. Pickering, "Immunization Policy Development in the United States: The Role of the Advisory Committee on Immunization Practices," *Annals of Internal Medicine* 150, no. 1 (January 2009): 45–49, doi:10.7326/0003-4819-150-1-200901060-00009.
48. CDC, "Rotavirus Vaccine for the Prevention of Rotavirus Gastroenteritis among Children: Recommendations of the Advisory Committee on Immunization Practices (ACIP)," *MMWR Recommendations and Reports* 48, no. RR-2 (1999), http://www.cdc.gov/mmwr/preview/mmwrhtml/00056669.htm.
49. CDC, "Vaccine Adverse Event Reporting System (VAERS)," 2013, http://www.cdc.gov/vaccinesafety/Activities/vaers.html; F. Varricchio et al., "Understanding Vaccine Safety Information from the Vaccine Adverse Event Reporting System," *Pediatric Infectious Disease Journal* 23, no. 4 (2004): 287–94.
50. CDC, "Withdrawal of Rotavirus Vaccine Recommendation," *MMWR Recommendations and Reports* 48, no. 43 (1999), http://www.cdc.gov/mmwr/preview/mmwrhtml/mm4843a5.htm; Schwartz, "The First Rotavirus Vaccine."
51. A. Allen, "No Good Deed Goes Unpunished," in *Vaccine: The Controversial Story of Medicine's Greatest Lifesaver* (New York: W. W. Norton, 2007), 294–326.
52. J. Baggs et al., "The Vaccine Safety Datalink: A Model for Monitoring Immunization Safety," *Pediatrics* 127, suppl. (2011): S45-53, doi:10.1542/peds.2010-1722H; CDC, "Vaccine Safety Datalink (VSD)," 2013, http://www.cdc.gov/vaccinesafety/Activities/vsd.html.
53. W. K. Yih et al., "Intussusception Risk after Rotavirus Vaccination in U.S. Infants," *New England Journal of Medicine* 370, no. 6 (2014): 503–12, doi:10.1056/NEJMoa1303164; E. S. Weintraub et al., "Risk of Intussusception after Monovalent Rotavirus Vaccination," *New England Journal of Medicine* 370, no. 6 (2014): 513–19, doi:10.1056/NEJMoa1311738.
54. C. Yen et al., "Trends in Intussusception Hospitalizations among US Infants before and after Implementation of the Rotavirus Vaccination Program, 2000–2009," *Journal of Infectious Diseases* 206, no. 1 (2012): 41–48, doi:10.1093/infdis/jis314.
55. B. R. Murphy et al., "Reappraisal of the Association of Intussusception with the Licensed Live Rotavirus Vaccine Challenges Initial Conclusions," *Journal of Infectious Diseases* 187, no. 8 (2003): 1301–8.
56. R. Desai et al., "Potential Intussusception Risk versus Benefits of Rotavirus Vaccination in the United States," *Pediatric Infectious Disease Journal* 32, no. 1 (2013): 1–7, doi:10.1097/INF.0b013e318270362c.

57. Janie Oyakawa, email to the author, March 12, 2014.
58. J. Oyakawa, "A Crunchy Mom's Reversal on Vaccinations," *The Mom of OZ* (blog), November 10, 2013, http://rubyslippersx3.blogspot.co.uk/2013/11/a-crunchy-moms-reversal-on-vaccinations.html.
59. A. Kennedy et al., "Confidence about Vaccines in the United States: Understanding Parents' Perceptions," *Health Affairs (Millwood)* 30, no. 6 (2011): 1151–59, doi:10.1377/hlthaff.2011.0396; M. Mason McCauley et al., "Exploring the Choice to Refuse or Delay Vaccines: A National Survey of Parents of 6- through 23-Month-Olds," *Academic Pediatrics* 12, no. 5 (2012): 375–83, doi:10.1016/j.acap.2012.06.007.
60. Paul Slovic, interview with the author, Eugene, OR, February 18, 2014.
61. D. Kahneman, "Of Two Minds: How Fast and Slow Thinking Shape Perception and Choice," *Scientific American*, June 15, 2012, http://www.scientificamerican.com/article/kahneman-excerpt-thinking-fast-and-slow; P. Slovic et al., "Risk as Analysis and Risk as Feelings: Some Thoughts about Affect, Reason, Risk, and Rationality," *Risk Analysis* 24, no. 2 (2004): 311–22, doi:10.1111/j.0272–4332.2004.00433.x.
62. P. Slovic, "Perception of Risk," *Science* 236, no. 4799 (1987): 280–85; L. K. Ball, G. Evans, and A. Bostrom, "Risky Business: Challenges in Vaccine Risk Communication," *Pediatrics* 101, no. 3 (1998): 453–58; M. Siddiqui, D. A. Salmon, and S. B. Omer, "Epidemiology of Vaccine Hesitancy in the United States," *Human Vaccines and Immunotherapeutics* 9, no. 12 (2013): 2643–48, doi:10.4161/hv.27243.
63. Slovic, interview with the author.
64. Slovic, "Perception of Risk"; D. M. Kahan et al., "Who Fears the HPV Vaccine, Who Doesn't, and Why? An Experimental Study of the Mechanisms of Cultural Cognition," *Law and Human Behavior* 34, no. 6 (2010): 501–16, doi:10.1007/s10979-009–9201–0.
65. Oyakawa, "A Crunchy Mom's Reversal."

CHAPTER 9. GETTING STARTED WITH SOLID FOODS

1. S. J. Fomon, "Infant Feeding in the 20th Century: Formula and Beikost," *Journal of Nutrition* 131 (2001): 409-20S.
2. K. G. Dewey and C. M. Chaparro, "Session 4: Mineral Metabolism and Body Composition Iron Status of Breast-Fed Infants," *Proceedings of the Nutrition Society* 66 (August 2007): 412–22, doi:10.1017/S002966510700568X.
3. C. J. Chantry, C. R. Howard, and P. Auinger, "Full Breastfeeding Duration and Risk for Iron Deficiency in U.S. Infants," *Breastfeeding Medicine* 2, no. 2 (2007): 63–73, doi:10.1089/bfm.2007.0002; E. B. Calvo, A. C. Galindo, and N. B. Aspres, "Iron Status in Exclusively Breast-Fed Infants," *Pediatrics* 90, no. 3 (1992): 375; D. Hopkins et al., "Infant Feeding in the Second 6 Months of Life Related to Iron Status: An Observational Study," *Archives of Disease in Childhood* 92 (2007): 850–54.

4. N. F. Krebs et al., "Zinc Supplementation during Lactation: Effects on Maternal Status and Milk Zinc Concentrations," *American Journal of Clinical Nutrition* 61, no. 5 (May 1, 1995): 1030–36.
5. Ibid.; M. A. Siimes, L. Salmenperä, and J. Perheentupa, "Exclusive Breast-Feeding for 9 Months: Risk of Iron Deficiency," *Journal of Pediatrics* 104, no. 2 (1984): 196–99.
6. N. F. Krebs et al., "Meat as a First Complementary Food for Breastfed Infants: Feasibility and Impact on Zinc Intake and Status," *Journal of Pediatric Gastroenterology and Nutrition* 42, no. 2 (2006): 207–14; L. Persson et al., "Are Weaning Foods Causing Impaired Iron and Zinc Status in 1-Year-Old Swedish Infants? A Cohort Study," *Acta Paediatrica* 87, no. 6 (1998): 618–22, doi:10.1111/j.1651–2227.1998.tb01518.x; R. G. Sezer et al., "Effect of Breastfeeding on Serum Zinc Levels and Growth in Healthy Infants," *Breastfeeding Medicine* 8 (2013): 159–63, doi:10.1089/bfm.2012.0014.
7. K. G. Dewey, *Guiding Principles for Complementary Feeding of the Breastfed Child* (Washington, DC: Pan American Health Organization, World Health Organization, 2003).
8. K. G. Dewey, "The Challenge of Meeting Nutrient Needs of Infants and Young Children during the Period of Complementary Feeding: An Evolutionary Perspective," *Journal of Nutrition* 143, no. 12 (2013): 2050–54, doi:10.3945/jn.113.182527.
9. World Health Organization (WHO), "Complementary Feeding," accessed November 1, 2013, http://www.who.int/nutrition/topics/complementary_feeding/en.
10. Sezer et al., "Effect of Breastfeeding on Serum Zinc"; R. D. Baker and F. R. Greer, "Diagnosis and Prevention of Iron Deficiency and Iron-Deficiency Anemia in Infants and Young Children (0–3 Years of Age)," *Pediatrics* 126 (2010): 1040–50, doi:10.1542/peds.2010–2576.
11. American Academy of Pediatrics (AAP) Section on Breastfeeding, "Breastfeeding and the Use of Human Milk," *Pediatrics* 129, no. 3 (2012): e827-41, doi:10.1542/peds.2011–3552; C. Agostoni et al., "Complementary Feeding: A Commentary by the ESPGHAN Committee on Nutrition," *Journal of Pediatric Gastroenterology and Nutrition* 46, no. 1 (2008): 99–110; European Food Safety Authority (EFSA) Panel on Dietetic Products, Nutrition and Allergies, "Scientific Opinion on the Appropriate Age for Introduction of Complementary Feeding of Infants," *European Food Safety Authority Journal* 7, no. 12 (2009): 2–38.
12. H. B. Clayton et al., "Prevalence and Reasons for Introducing Infants Early to Solid Foods: Variations by Milk Feeding Type," *Pediatrics* 131, no. 4 (2013): e1108-14, doi:10.1542/peds.2012–2265.
13. A. A. Kuo et al., "Introduction of Solid Food to Young Infants," *Maternal and Child Health Journal* 15, no. 8 (2010): 1185–94, doi:10.1007/s10995-010–0669–5.
14. WHO, "Complementary Feeding"; Dewey, "Guiding Principles for Complementary Feeding."

15. AAP Section on Breastfeeding, "Breastfeeding and the Use of Human Milk"; U.K. National Health Service, "Your Baby's First Solid Foods," May 23, 2014, http://www.nhs.uk/conditions/pregnancy-and-baby/pages/solid-foods-weaning.aspx#close.
16. Valerie Wheat, email to the author, November 23, 2013.
17. M. Fewtrell et al., "Six Months of Exclusive Breast Feeding: How Good Is the Evidence?" *BMJ* 342 (2011): 209–12, doi:10.1136/bmj.c5955.
18. R. J. Cohen et al., "Effects of Age of Introduction of Complementary Foods on Infant Breast Milk Intake, Total Energy Intake, and Growth: A Randomised Intervention Study in Honduras," *Lancet* 344, no. 8918 (1994): 288–93, doi:10.1016/S0140-6736(94)91337-4; K. G. Dewey et al., "Age of Introduction of Complementary Foods and Growth of Term, Low-Birth-Weight, Breast-Fed Infants: A Randomized Intervention Study in Honduras," *American Journal of Clinical Nutrition* 69, no. 4 (1999): 679–86.
19. O. H. Jonsdottir et al., "Exclusive Breastfeeding for 4 versus 6 Months and Growth in Early Childhood," *Acta Paediatrica* 103, no. 1 (2013): 105–11, doi:10.1111/apa.12433; O. H. Jonsdottir et al., "Timing of the Introduction of Complementary Foods in Infancy: A Randomized Controlled Trial," *Pediatrics* 130, no. 6 (December 1, 2012): 1038–45, doi:10.1542/peds.2011–3838.
20. M. S. Kramer and R. Kakuma, "Optimal Duration of Exclusive Breastfeeding," *Cochrane Database of Systematic Reviews* 8 (2012): CD003517, doi:10.1002/14651858.CD003517.pub2.
21. Ibid.
22. Ibid.
23. B. M. Popkin et al., "Breast-Feeding and Diarrheal Morbidity," *Pediatrics* 86, no. 6 (1990): 874–82.
24. M. S. Kramer et al., "Infant Growth and Health Outcomes Associated with 3 Compared with 6 Mo of Exclusive Breastfeeding," *American Journal of Clinical Nutrition* 78, no. 2 (2003): 291–95.
25. Kramer and Kakuma, "Optimal Duration of Exclusive Breastfeeding."
26. Cohen et al., "Effects of Age of Introduction of Complementary Foods"; Dewey et al., "Age of Introduction of Complementary Foods."
27. M. A. Quigley, Y. J. Kelly, and A. Sacker, "Infant Feeding, Solid Foods and Hospitalisation in the First 8 Months after Birth," *Archives of Disease in Childhood* 94, no. 2 (February 1, 2009): 148–50, doi:10.1136/adc.2008.146126.
28. J. F. Ludvigsson and A. Fasano, "Timing of Introduction of Gluten and Celiac Disease Risk," *Annals of Nutrition and Metabolism* 60, suppl. 2 (2012): 22–29, doi:10.1159/000335335; J. M. Norris et al., "Risk of Celiac Disease Autoimmunity and Timing of Gluten Introduction in the Diet of Infants at Increased Risk of Disease," *JAMA* 293, no. 19 (2005): 2343–51, doi:10.1001/jama.293.19.2343; K. Størdal, R. A. White, and M. Eggesbø, "Early Feeding and Risk of Celiac Disease in a Prospective Birth Cohort," *Pediatrics*, October 7, 2013, peds.2013–1752, doi:10.1542/peds.2013–1752.

29. B. Frederiksen et al., "Infant Exposures and Development of Type 1 Diabetes Mellitus: The Diabetes Autoimmunity Study in the Young (DAISY)," *JAMA Pediatrics* 167, no. 9 (2013): 808–15, doi:10.1001/jamapediatrics.2013.317; J. M. Norris et al., "Timing of Initial Cereal Exposure in Infancy and Risk of Islet Autoimmunity," *JAMA* 290, no. 13 (2003): 1713–20, doi:10.1001/jama.290.13.1713; A. G. Ziegler et al., "Early Infant Feeding and Risk of Developing Type 1 Diabetes–Associated Autoantibodies," *JAMA* 290, no. 13 (2003): 1721–28, doi:10.1001/jama.290.13.1721.
30. K. E. C. Grimshaw et al., "Introduction of Complementary Foods and the Relationship to Food Allergy," *Pediatrics* 132, no. 6 (2013): 1529–38; B. E. P. Snijders et al., "Age at First Introduction of Cow Milk Products and Other Food Products in Relation to Infant Atopic Manifestations in the First 2 Years of Life: The KOALA Birth Cohort Study," *Pediatrics* 122, no. 1 (2008): e115-22, doi:10.1542/peds.2007–1651; A. Zutavern et al., "Timing of Solid Food Introduction in Relation to Eczema, Asthma, Allergic Rhinitis, and Food and Inhalant Sensitization at the Age of 6 Years: Results from the Prospective Birth Cohort Study LISA," *Pediatrics* 121, no. 1 (2008): e44-52, doi:10.1542/peds.2006–3553.
31. J. A. Poole et al., "Timing of Initial Exposure to Cereal Grains and the Risk of Wheat Allergy," *Pediatrics* 117, no. 6 (2006): 2175–82, doi:10.1542/peds.2005–1803.
32. J. J. Koplin et al., "Can Early Introduction of Egg Prevent Egg Allergy in Infants? A Population-Based Study," *Journal of Allergy and Clinical Immunology* 126, no. 4 (2010): 807–13, doi:10.1016/j.jaci.2010.07.028.
33. G. Du Toit et al., "Early Consumption of Peanuts in Infancy Is Associated with a Low Prevalence of Peanut Allergy," *Journal of Allergy and Clinical Immunology* 122, no. 5 (2008): 984–91, doi:10.1016/j.jaci.2008.08.039.
34. LEAP (Learning Early about Peanut Allergy), "The LEAP Study: A Clinical Trial on the Prevention of Peanut Allergy," accessed November 14, 2013, http://www.leapstudy.co.uk/index.html.
35. K. D. Jackson, L. D. Howie, and L. J. Akinbami, *Trends in Allergic Conditions among Children: United States, 1997–2011*, NCHS Data Brief, no. 121 (Hyattsville, MD: National Center for Health Statistics, 2013).
36. S. H. Sicherer, A. Muñoz-Furlong, and H. A. Sampson, "Prevalence of Peanut and Tree Nut Allergy in the United States Determined by Means of a Random Digit Dial Telephone Survey: A 5-Year Follow-up Study," *Journal of Allergy and Clinical Immunology* 112, no. 6 (2003): 1203–7, doi:10.1016/S0091.
37. S. L. Prescott et al., "The Importance of Early Complementary Feeding in the Development of Oral Tolerance: Concerns and Controversies," *Pediatric Allergy and Immunology: Official Publication of the European Society of Pediatric Allergy and Immunology* 19, no. 5 (August 2008): 375–80, doi:10.1111/j.1399–3038.2008.00718.x.
38. F. R. Greer, S. H. Sicherer, and A. W. Burks, "Effects of Early Nutritional Interventions on the Development of Atopic Disease in Infants and Children: The

Role of Maternal Dietary Restriction, Breastfeeding, Timing of Introduction of Complementary Foods, and Hydrolyzed Formulas," *Pediatrics* 121, no. 1 (2008): 183–91, doi:10.1542/peds.2007-3022.

39. A. Høst et al., "Dietary Prevention of Allergic Diseases in Infants and Small Children," *Pediatric Allergy and Immunology* 19, no. 1 (2008): 1–4, doi:10.1111/j.1399-3038.2007.00680.x; S. L. Prescott and M. L. K. Tang, "The Australasian Society of Clinical Immunology and Allergy Position Statement: Summary of Allergy Prevention in Children," *Medical Journal of Australia* 182, no. 9 (2005): 464–67; E. S. Chan, C. Cummings, and Canadian Paediatric Society, Community Paediatrics Committee, Allergy Section, "Dietary Exposures and Allergy Prevention in High-Risk Infants," *Paediatrics and Child Health* 18, no. 10 (2013): 545–49.
40. J. van Odijk et al., "Breastfeeding and Allergic Disease: A Multidisciplinary Review of the Literature (1966–2001) on the Mode of Early Feeding in Infancy and Its Impact on Later Atopic Manifestations," *Allergy* 58, no. 9 (2003): 833–43, doi:10.1034/j.1398-9995.2003.00264.x; B. Laubereau et al., "Effect of Breast-Feeding on the Development of Atopic Dermatitis during the First 3 Years of Life: Results from the GINI-Birth Cohort Study," *Journal of Pediatrics* 144, no. 5 (2004): 602–7, doi:10.1016/j.jpeds.2003.12.029.
41. M. S. Kramer et al., "Effect of Prolonged and Exclusive Breast Feeding on Risk of Allergy and Asthma: Cluster Randomised Trial," *BMJ* 335, no. 7624 (2007): 815, doi:10.1136/bmj.39304.464016.AE; M. C. Matheson et al., "Breast-Feeding and Atopic Disease: A Cohort Study from Childhood to Middle Age," *Journal of Allergy and Clinical Immunology* 120, no. 5 (2007): 1051–57, doi:10.1016/j.jaci.2007.06.030; B. I. Nwaru et al., "Breastfeeding and Introduction of Complementary Foods during Infancy in Relation to the Risk of Asthma and Atopic Diseases up to 10 Years," *Clinical and Experimental Allergy* 43, no. 11 (2013): 1263–73, doi:10.1111/cea.12180.
42. Grimshaw et al., "Introduction of Complementary Foods."
43. CDC, "Breastfeeding: Breastfeeding Report Card—United States, 2012," 2013, http://www.cdc.gov/breastfeeding/data/reportcard/reportcard2012.htm.
44. Quigley, Kelly, and Sacker, "Infant Feeding."
45. M. P. W. Platt, "Demand Weaning: Infants' Answer to Professionals' Dilemmas," *Archives of Disease in Childhood* 94, no. 2 (2009): 79–80, doi:10.1136/adc.2008.150011.
46. E. Satter, "Feeding Your Older Baby: Oral-Motor Development," in *Child of Mine: Feeding with Love and Good Sense* (Boulder, CO: Bull Publishing, 2000), 267–69.
47. B. R. Carruth and J. D. Skinner, "Feeding Behaviors and Other Motor Development in Healthy Children (2–24 Months)," *Journal of the American College of Nutrition* 21, no. 2 (2002): 88–96, doi:10.1080/07315724.2002.10719199.
48. Ibid.; B. R. Carruth et al., "Developmental Milestones and Self-Feeding Behaviors in Infants and Toddlers," *Journal of the American Dietetic Association* 104, suppl. 1 (2004): 51–56, doi:10.1016/j.jada.2003.10.019.

49. E. G. Gisel, "Effect of Food Texture on the Development of Chewing of Children between Six Months and Two Years of Age," *Developmental Medicine and Child Neurology* 33, no. 1 (1991): 69–79, doi:10.1111/j.1469-8749.1991.tb14786.x.
50. Carruth et al., "Developmental Milestones."
51. K. Northstone et al., "The Effect of Age of Introduction to Lumpy Solids on Foods Eaten and Reported Feeding Difficulties at 6 and 15 Months," *Journal of Human Nutrition and Dietetics* 14, no. 1 (2001): 43–54, doi:10.1046/j.1365-277x.2001.00264.x; H. Coulthard, G. Harris, and P. Emmett, "Delayed Introduction of Lumpy Foods to Children during the Complementary Feeding Period Affects Child's Food Acceptance and Feeding at 7 Years of Age," *Maternal and Child Nutrition* 5, no. 1 (2009): 75–85, doi:10.1111/j.1740-8709.2008.00153.x.
52. M. M. Black and F. E. Aboud, "Responsive Feeding Is Embedded in a Theoretical Framework of Responsive Parenting," *Journal of Nutrition* 141, no. 3 (2011): 490–94, doi:10.3945/jn.110.129973.
53. M. K. Fox et al., "Relationship between Portion Size and Energy Intake among Infants and Toddlers: Evidence of Self-Regulation," *Journal of the American Dietetic Association* 106, no. 1, suppl. (January 2006): 77–83, doi:10.1016/j.jada.2005.09.039.
54. C. Schwartz et al., "Development of Healthy Eating Habits Early in Life: Review of Recent Evidence and Selected Guidelines," *Appetite* 57, no. 3 (2011): 796–807, doi:10.1016/j.appet.2011.05.316; A. K. Ventura and L. L. Birch, "Does Parenting Affect Children's Eating and Weight Status?" *International Journal of Behavioral Nutrition and Physical Activity* 5 (2008): 15, doi:10.1186/1479-5868-5-15.
55. C.V. Farrow and J. Blissett, "Controlling Feeding Practices: Cause or Consequence of Early Child Weight?" *Pediatrics* 121, no. 1 (2008): e164-69, doi:10.1542/peds.2006-3437; C. Farrow and J. Blissett, "Does Maternal Control during Feeding Moderate Early Infant Weight Gain?" *Pediatrics* 118, no. 2 (2006): e293-98, doi:10.1542/peds.2005-2919.
56. Black and Aboud, "Responsive Feeding."
57. G. Rapley, "Baby-Led Weaning Pamphlet," accessed December 29, 2013, http://www.rapleyweaning.com.
58. A. Brown and M. Lee, "A Descriptive Study Investigating the Use and Nature of Baby-Led Weaning in a UK Sample of Mothers," *Maternal and Child Nutrition* 7, no. 1 (2011): 34–47, doi:10.1111/j.1740-8709.2010.00243.x; A. Brown and M. Lee, "Maternal Control of Child Feeding during the Weaning Period: Differences between Mothers Following a Baby-Led or Standard Weaning Approach," *Maternal and Child Health Journal* 15, no. 8 (November 2011): 1265–71, doi:10.1007/s10995-010-0678-4.
59. C. M. Wright et al., "Is Baby-Led Weaning Feasible? When Do Babies First Reach Out for and Eat Finger Foods?" *Maternal and Child Nutrition* 7, no. 1 (2011): 27–33, doi:10.1111/j.1740-8709.2010.00274.x.
60. Ibid.
61. E. Townsend and N. J. Pitchford, "Baby Knows Best? The Impact of Wean-

ing Style on Food Preferences and Body Mass Index in Early Childhood in a Case-Controlled Sample," *BMJ Open* 2, no. 1 (2012): e000298, doi:10.1136/bmjopen-2011-000298.

CHAPTER 10. EAT, GROW, AND LEARN

1. Eve F., post on *Natural Mother Magazine* Facebook page asking for advice about how to start solid foods, accessed October 20, 2013, https://www.facebook.com/NaturalMotherMagazine.
2. S. J. Fomon, "Infant Feeding in the 20th Century: Formula and Beikost," *Journal of Nutrition* 131 (2001): 409-20S.
3. A. M. Siega-Riz et al., "Food Consumption Patterns of Infants and Toddlers: Where Are We Now?" *Journal of the American Dietetic Association* 110 (2010): S38-51, doi:10.1016/j.jada.2010.09.001.
4. B. S. Vitta and K. G. Dewey, *Identifying Micronutrient Gaps in the Diets of Breastfed 6–11 Mo-Old Infants in Bangladesh, Ethiopia, and Vietnam Using Linear Programming* (Washington, DC: Alive and Thrive, 2012).
5. K. G. Dewey, "The Challenge of Meeting Nutrient Needs of Infants and Young Children during the Period of Complementary Feeding: An Evolutionary Perspective," *Journal of Nutrition* 143, no. 12 (2013): 2050–54, doi:10.3945/jn.113.182527.
6. Pan American Health Organization (PAHO), World Health Organization (WHO), "Guiding Principles for Complementary Feeding of the Breastfed Child," World Health Organization, 2003, http://www.who.int/maternal_child_adolescent/documents/a85622/en.
7. R. Yip et al., "Declining Prevalence of Anemia among Low-Income Children in the United States," *JAMA* 258, no. 12 (1987): 1619–23.
8. Dewey, "The Challenge of Meeting Nutrient Needs."
9. G. H. Pelto, Y. Zhang, and J. Habicht, "Premastication: The Second Arm of Infant and Young Child Feeding for Health and Survival?" *Maternal and Child Nutrition* 6, no. 1 (2010): 4–18, doi:10.1111/j.1740–8709.2009.00200.x.
10. L. Cordain et al., "Fatty Acid Analysis of Wild Ruminant Tissues: Evolutionary Implications or Reducing Diet-Related Chronic Disease," *European Journal of Clinical Nutrition* 56, no. 3 (2002): 181–91.
11. E. E. Birch et al., "The DIAMOND (DHA Intake And Measurement Of Neural Development) Study: A Double-Masked, Randomized Controlled Clinical Trial of the Maturation of Infant Visual Acuity as a Function of the Dietary Level of Docosahexaenoic Acid," *American Journal of Clinical Nutrition* 91, no. 4 (2010): 848–59, doi:10.3945/ajcn.2009.28557.
12. C. Campoy et al., "Omega 3 Fatty Acids on Child Growth, Visual Acuity and Neurodevelopment," *British Journal of Nutrition* 107, suppl. 2 (2012): S85-106, doi:10.1017/S0007114512001493.
13. G. A. Olaya, M. Lawson, and M. S. Fewtrell, "Efficacy and Safety of New Complementary Feeding Guidelines with an Emphasis on Red Meat Consumption:

A Randomized Trial in Bogota, Colombia," *American Journal of Clinical Nutrition* 98, no. 4 (October 2013): 983–93, doi:10.3945/ajcn.112.053595; N. F. Krebs et al., "Comparison of Complementary Feeding Strategies to Meet Zinc Requirements of Older Breastfed Infants," *American Journal of Clinical Nutrition* 96, no. 1 (2012): 30–35, doi:10.3945/ajcn.112.036046.

14. K. M. Hambidge et al., "Evaluation of Meat as a First Complementary Food for Breastfed Infants: Impact on Iron Intake," *Nutrition Reviews* 69 (2011): S57-63.
15. L. Davidsson et al., "Iron Bioavailability in Infants from an Infant Cereal Fortified with Ferric Pyrophosphate or Ferrous Fumarate," *American Journal of Clinical Nutrition* 71, no. 6 (2000): 1597–1602; Joint FAO/WHO Expert Consultation on Human Vitamin and Mineral Requirements and WHO Department of Nutrition for Health and Development, *Vitamin and Mineral Requirements in Human Nutrition*, 2nd ed. (Geneva: World Health Organization, 2005).
16. J. D. Cook and E. R. Monsen, "Food Iron Absorption in Human Subjects: III. Comparison of the Effect of Animal Proteins on Nonheme Iron Absorption," *American Journal of Clinical Nutrition* 29, no. 8 (1976): 859–67.
17. Siega-Riz et al., "Food Consumption Patterns."
18. American Academy of Pediatrics (AAP), "Switching to Solid Foods," HealthyChildren.org, accessed November 11, 2013, http://www.healthychildren.org/English/ages-stages/baby/feeding-nutrition/Pages/Switching-To-Solid-Foods.aspx.
19. J. Morgan, A. Taylor, and M. Fewtrell, "Meat Consumption Is Positively Associated with Psychomotor Outcome in Children up to 24 Months of Age," *Journal of Pediatric Gastroenterology and Nutrition* 39, no. 5 (November 2004): 493–98.
20. N. F. Krebs et al., "Meat as a First Complementary Food for Breastfed Infants: Feasibility and Impact on Zinc Intake and Status," *Journal of Pediatric Gastroenterology and Nutrition* 42, no. 2 (2006): 207–14.
21. S. M. Innis, "Dietary Omega 3 Fatty Acids and the Developing Brain," *Brain Research* 1237 (2008): 35–43, doi:10.1016/j.brainres.2008.08.078; C. A. Daley et al., "A Review of Fatty Acid Profiles and Antioxidant Content in Grass-Fed and Grain-Fed Beef," *Nutrition Journal* 9 (2010): 10, doi:10.1186/1475–2891–9–10.
22. Hambidge et al., "Evaluation of Meat."
23. L. Hallberg and L. Hulthén, "Prediction of Dietary Iron Absorption: An Algorithm for Calculating Absorption and Bioavailability of Dietary Iron," *American Journal of Clinical Nutrition* 71, no. 5 (2000): 1147–60; R. F. Hurrell et al., "Iron Absorption in Humans: Bovine Serum Albumin Compared with Beef Muscle and Egg White," *American Journal of Clinical Nutrition* 47, no. 1 (1988): 102–7.
24. M. Makrides et al., "Nutritional Effect of Including Egg Yolk in the Weaning Diet of Breast-Fed and Formula-Fed Infants: A Randomized Controlled Trial," *American Journal of Clinical Nutrition* 75, no. 6 (2002): 1084–92.
25. Ibid.

26. U.S. Food and Drug Administration, "Metals: Fish—What Pregnant Women and Parents Should Know," 2014, http://www.fda.gov/Food/Foodborne IllnessContaminants/Metals/ucm393070.htm.
27. I. Fraeye et al., "Dietary Enrichment of Eggs with Omega-3 Fatty Acids: A Review," *Food Research International* 48, no. 2 (2012): 961–69, doi:10.1016/j.foodres.2012.03.014.
28. D. R. Hoffman et al., "Maturation of Visual Acuity Is Accelerated in Breast-Fed Term Infants Fed Baby Food Containing DHA-Enriched Egg Yolk," *Journal of Nutrition* 134, no. 9 (2004): 2307–13.
29. A. P. Simopoulos and N. Salem, "Egg Yolk as a Source of Long-Chain Polyunsaturated Fatty Acids in Infant Feeding," *American Journal of Clinical Nutrition* 55, no. 2 (1992): 411–14; H. D. Karsten et al., "Vitamins A, E and Fatty Acid Composition of the Eggs of Caged Hens and Pastured Hens," *Renewable Agriculture and Food Systems* 25, special issue 1 (2010): 45–54, doi:10.1017/S1742170509990214.
30. USDA Food Safety and Inspection Service, "Meat and Poultry Labeling Terms," accessed January 21, 2014, http://www.fsis.usda.gov/wps/portal/fsis/topics/food-safety-education/get-answers/food-safety-fact-sheets/food-labeling/meat-and-poultry-labeling-terms/meat-and-poultry-labeling-terms.
31. Makrides et al., "Nutritional Effect of Including Egg Yolk."
32. S. H. Zeisel and K. A. da Costa, "Choline: An Essential Nutrient for Public Health," *Nutrition Reviews* 67, no. 11 (2009): 615–23.
33. E. E. Ziegler, "Consumption of Cow's Milk as a Cause of Iron Deficiency in Infants and Toddlers," *Nutrition Reviews* 69 (2011): S37-42, doi:10.1111/j.1753-4887.2011.00431.x.
34. C. Agostoni et al., "Complementary Feeding: A Commentary by the ESPGHAN Committee on Nutrition," *Journal of Pediatric Gastroenterology and Nutrition* 46, no. 1 (2008): 99–110.
35. PAHO, WHO, "Guiding Principles for Complementary Feeding."
36. P. C. Dagnelie and W. A. van Staveren, "Macrobiotic Nutrition and Child Health: Results of a Population-Based, Mixed-Longitudinal Cohort Study in The Netherlands," *American Journal of Clinical Nutrition* 59, no. 5 (May 1994): 1187-96S.
37. M. Van Winckel et al., "Vegetarian Infant and Child Nutrition," *European Journal of Pediatrics* 170, no. 12 (December 1, 2011): 1489–94, doi:10.1007/s00431-011-1547-x.
38. E. E. Ziegler, S. E. Nelson, and J. M. Jeter, "Iron Status of Breastfed Infants Is Improved Equally by Medicinal Iron and Iron-Fortified Cereal," *American Journal of Clinical Nutrition* 90 (July 2009): 76–87, doi:10.3945/ajcn.2008.27350.
39. J. Mennella and G. Beauchamp, "Mothers' Milk Enhances the Acceptance of Cereal during Weaning," *Pediatric Research* 41, no. 2 (1997): 188–92.
40. FAO/WHO, *Vitamin and Mineral Requirements*.
41. M. B. Zimmermann and R. F. Hurrell, "Nutritional Iron Deficiency," *Lancet* 370, no. 9586 (2007): 511–20, doi:10.1016/S0140-6736(07)61235-5.

42. J. D. Cook et al., "The Influence of Different Cereal Grains on Iron Absorption from Infant Cereal Foods," *American Journal of Clinical Nutrition* 65, no. 4 (1997): 964–69.
43. N. Roos et al., "Screening for Anti-Nutritional Compounds in Complementary Foods and Food Aid Products for Infants and Young Children: Anti-Nutritional Compounds in Complementary Foods and Products," *Maternal and Child Nutrition* 9 (2013): 47–71, doi:10.1111/j.1740–8709.2012.00449.x.
44. N. F. Krebs et al., "Effects of Different Complementary Feeding Regimens on Iron Status and Enteric Microbiota in Breastfed Infants," *Journal of Pediatrics* 163, no. 2 (2013): 416–23, doi:10.1016/j.jpeds.2013.01.024.
45. L. A. Perlas and R. S. Gibson, "Household Dietary Strategies to Enhance the Content and Bioavailability of Iron, Zinc and Calcium of Selected Rice- and Maize-Based Philippine Complementary Foods," *Maternal and Child Nutrition* 1, no. 4 (October 2005): 263–73, doi:10.1111/j.1740–8709.2005.00037.x.
46. R. S. Gibson et al., "A Review of Phytate, Iron, Zinc, and Calcium Concentrations in Plant-Based Complementary Foods Used in Low-Income Countries and Implications for Bioavailability," *Food and Nutrition Bulletin* 31, no. 2 (2010): 134–46.
47. B. Hadorn et al., "Quantitative Assessment of Exocrine Pancreatic Function in Infants and Children," *Journal of Pediatrics* 73, no. 1 (1968): 39–50, doi:10.1016/S0022-3476(68)80037-X; G. Zoppi et al., "Exocrine Pancreas Function in Premature and Full Term Neonates," *Pediatric Research* 6, no. 12 (1972): 880–86, doi:10.1203/00006450–197212000–00005.
48. B. De Vizia et al., "Digestibility of Starches in Infants and Children," *Journal of Pediatrics* 86, no. 1 (1975): 50–55, doi:10.1016/S0022-3476(75)80703-7.
49. Fomon, "Infant Feeding."
50. G. H. Pelto, E. Levitt, and L. Thairu, "Improving Feeding Practices: Current Patterns, Common Constraints, and the Design of Interventions," *Food and Nutrition Bulletin* 24, no. 1 (2003): 45–82.
51. M. A. Rossiter et al., "Amylase Content of Mixed Saliva in Children," *Acta Paediatrica* 63, no. 3 (1974): 389–92, doi:10.1111/j.1651–2227.1974.tb04815.x; G. P. Sevenhuysen, C. Holodinsky, and C. Dawes, "Development of Salivary Alpha-Amylase in Infants from Birth to 5 Months," *American Journal of Clinical Nutrition* 39, no. 4 (1984): 584–88.
52. R. D. Murray et al., "The Contribution of Salivary Amylase to Glucose Polymer Hydrolysis in Premature Infants," *Pediatric Research* 20, no. 2 (1986): 186–91, doi:10.1203/00006450–198602000–00019; J. L. Rosenblum, C. L. Irwin, and D. H. Alpers, "Starch and Glucose Oligosaccharides Protect Salivary-Type Amylase Activity at Acid pH," *American Journal of Physiology* 254, no. 5 (1988): G775-80.
53. K. M. Shahani, A. J. Kwan, and B. A. Friend, "Role and Significance of Enzymes in Human Milk," *American Journal of Clinical Nutrition* 33, no. 8 (August 1, 1980): 1861–68.

54. J. B. Jones, N. R. Mehta, and M. Hamosh, "Alpha-Amylase in Preterm Human Milk," *Journal of Pediatric Gastroenterology and Nutrition* 1, no. 1 (1982): 43–48.
55. T. Lindberg and G. Skude, "Amylase in Human Milk," *Pediatrics* 70, no. 2 (1982): 235–38; L. A. Heitlinger et al., "Mammary Amylase: A Possible Alternate Pathway of Carbohydrate Digestion in Infancy," *Pediatric Research* 17, no. 1 (1983): 15–18, doi:10.1203/00006450-198301000-00003.
56. Jones, Mehta, and Hamosh, "Alpha-Amylase in Preterm Human Milk."
57. P. C. Lee et al., "Glucoamylase Activity in Infants and Children: Normal Values and Relationship to Symptoms and Histological Findings," *Journal of Pediatric Gastroenterology and Nutrition* 39, no. 2 (2004): 161–65; E. Lebenthal and P. C. Lee, "Glucoamylase and Disaccharidase Activities in Normal Subjects and in Patients with Mucosal Injury of the Small Intestine," *Journal of Pediatrics* 97, no. 3 (1980): 389–93, doi:10.1016/S0022-3476(80)80187-9.
58. R. J. Shulman et al., "Utilization of Dietary Cereal by Young Infants," *Journal of Pediatrics* 103, no. 1 (1983): 23–28, doi:10.1016/S0022-3476(83)80769-0; M. T. Christian et al., "Modeling 13C Breath Curves to Determine Site and Extent of Starch Digestion and Fermentation in Infants," *Journal of Pediatric Gastroenterology* 34, no. 2 (2002): 158–64.
59. J. M. W. Wong et al., "Colonic Health: Fermentation and Short Chain Fatty Acids," *Journal of Clinical Gastroenterology* 40, no. 3 (2006): 235–43.
60. M. T. Christian et al., "Starch Fermentation by Faecal Bacteria of Infants, Toddlers and Adults: Importance for Energy Salvage," *European Journal of Clinical Nutrition* 57, no. 11 (2003): 1486–91, doi:10.1038/sj.ejcn.1601715.
61. J. Scheiwiller et al., "Human Faecal Microbiota Develops the Ability to Degrade Type 3 Resistant Starch during Weaning," *Journal of Pediatric Gastroenterology and Nutrition* 43, no. 5 (2006): 584–91.
62. FAO/WHO, *Vitamin and Mineral Requirements.*
63. C. Schwartz et al., "Development of Healthy Eating Habits Early in Life: Review of Recent Evidence and Selected Guidelines," *Appetite* 57, no. 3 (2011): 796–807, doi:10.1016/j.appet.2011.05.316.
64. J. A. Mennella et al., "Variety Is the Spice of Life: Strategies for Promoting Fruit and Vegetable Acceptance during Infancy," *Physiology and Behavior* 94 (April 22, 2008): 29–38, doi:10.1016/j.physbeh.2007.11.014.
65. S. A. Sullivan and L. L. Birch, "Infant Dietary Experience and Acceptance of Solid Foods," *Pediatrics* 93 (February 1994): 271–77; A. S. Maier et al., "Effects of Repeated Exposure on Acceptance of Initially Disliked Vegetables in 7-Month-Old Infants," *Food Quality and Preference* 18 (2007): 1023–32.
66. C. J. Gerrish and J. A. Mennella, "Flavor Variety Enhances Food Acceptance in Formula-Fed Infants," *American Journal of Clinical Nutrition* 73, no. 6 (2001): 1080–85.
67. Siega-Riz et al., "Food Consumption Patterns."
68. Schwartz et al., "Development of Healthy Eating."

69. L. L. Birch, "Development of Food Preferences," *Annual Review of Nutrition* 19, no. 1 (1999): 41–62, doi:10.1146/annurev.nutr.19.1.41.
70. M. G. Tanzi and M. P. Gabay, "Association between Honey Consumption and Infant Botulism," *Pharmacotherapy: Journal of Human Pharmacology and Drug Therapy* 22, no. 11 (2002): 1479–83, doi:10.1592/phco.22.16.1479.33696.
71. J. M. Geleijnse et al., "Long-Term Effects of Neonatal Sodium Restriction on Blood Pressure," *Hypertension* 29, no. 4 (April 1, 1997): 913–17, doi:10.1161/01.HYP.29.4.913.

APPENDIXES

A. ARE THE INGREDIENTS IN THE NEWBORN VITAMIN K SHOT SAFE?

1. Centers for Disease Control and Prevention (CDC), "Notes from the Field: Late Vitamin K Deficiency Bleeding in Infants Whose Parents Declined Vitamin K Prophylaxis—Tennessee, 2013," *Morbidity and Mortality Weekly Report (MMWR)* 62, no. 45 (2013): 901–2.
2. Merck Manuals, "Drug Absorption," *Merck Manual for Healthcare Professionals*, 2014, http://www.merckmanuals.com/professional/clinical_pharmacology/pharmacokinetics/drug_absorption.html.
3. N. Shehab et al., "Exposure to the Pharmaceutical Excipients Benzyl Alcohol and Propylene Glycol among Critically Ill Neonates," *Pediatric Critical Care Medicine* 10, no. 2 (2009): 256–59, doi:10.1097/PCC.0b013e31819a383c; S. P. Nordt and L. E. Vivero, "Pharmaceutical Additives," in *Goldfrank's Toxicologic Emergencies*, 9th ed., ed. L. S. Nelson et al. (New York: McGraw-Hill, 2011).
4. H. D. Goff, "Colloidal Aspects of Ice Cream: A Review," *International Dairy Journal* 7, no. 6–7 (1997): 363–73, doi:10.1016/S0958-6946(97)00040-X.
5. V. Lorch et al., "Unusual Syndrome among Premature Infants: Association with a New Intravenous Vitamin E Product," *Pediatrics* 75, no. 3 (1985): 598; S. C. Smolinske, CRC *Handbook of Food, Drug, and Cosmetic Excipients* (Boca Raton, FL: CRC Press, 1992).
6. Smolinske, CRC *Handbook*.
7. U.S. National Library of Medicine, "Sodium Acetate Anhydrous Injection, Solution, Concentrate," *DailyMed*, accessed April 9, 2014, http://dailymed.nlm.nih.gov/dailymed/drugInfo.cfm?setid=c91ede49-c7cc-47b7-0f93-e6841dd5916a.
8. Hospira Inc., "Vitamin K1 Injection Label," accessed April 8, 2014, http://www.hospira.com/products_and_services/drugs/VITAMIN_K1_NEONATE.
9. R. J. Mitkus et al., "Updated Aluminum Pharmacokinetics Following Infant Exposures through Diet and Vaccination," *Vaccine* 29, no. 51 (2011): 9538–43, doi:10.1016/j.vaccine.2011.09.124.
10. National Center for Biotechnology Information, "Acetic Acid," *PubChem*, accessed April 9, 2014, http://pubchem.ncbi.nlm.nih.gov/summary/summary.cgi?cid=176.
11. Nordt and Vivero, "Pharmaceutical Additives"; American Academy of Pediatrics

(AAP), "'Inactive' Ingredients in Pharmaceutical Products: Update (Subject Review)," *Pediatrics* 99, no. 2 (1997): 268–78, doi:10.1542/peds.99.2.268; Shehab et al., "Exposure to the Pharmaceutical Excipients."

12. D. L. Riegert-Johnson and G. W. Volcheck, "The Incidence of Anaphylaxis Following Intravenous Phytonadione (Vitamin K1): A 5-Year Retrospective Review," *Annals of Allergy, Asthma, and Immunology* 89, no. 4 (2002): 400–406, doi:10.1016/S1081-1206(10)62042-X.
13. U.S. National Library of Medicine, "Dextrose (Dextrose Monohydrate), Injection, Solution," *DailyMed*, accessed April 8, 2014, http://dailymed.nlm.nih.gov/dailymed/drugInfo.cfm?setid=a7b4d5bf-472b-4f7d-b8c5-363eda02dfad.
14. Roche Products Limited, "Konokion MM Patient Information Leaflet," 2013, http://www.medicines.org.uk/emc/medicine/1699/SPC; F. R. Greer et al., "A New Mixed Micellar Preparation for Oral Vitamin K Prophylaxis: Randomised Controlled Comparison with an Intramuscular Formulation in Breast Fed Infants," *Archives of Disease in Childhood* 79, no. 4 (1998): 300–305, doi:10.1136/adc.79.4.300; G. Schubiger et al., "Prevention of Vitamin K Deficiency Bleeding with Oral Mixed Micellar Phylloquinone: Results of a 6-Year Surveillance in Switzerland," *European Journal of Pediatrics* 162, no. 12 (December 2003): 885–88, doi:10.1007/s00431-003-1327-3.
15. E. Hey, "Vitamin K: What, Why, and When," *Archives of Disease in Childhood: Fetal and Neonatal Edition* 88, no. 2 (2003): F80-83, doi:10.1136/fn.88.2.F80.

B. WHY IS THE HEPATITIS B VACCINE RECOMMENDED AT BIRTH?

1. Centers for Disease Control and Prevention (CDC), "A Comprehensive Immunization Strategy to Eliminate Transmission of Hepatitis B Virus in the United States," *MMWR Recommendations and Reports* 54, no. RR-16 (2005): 1–33.
2. C. W. Shepard et al., "Epidemiology of Hepatitis B and Hepatitis B Virus Infection in United States Children," *Pediatric Infectious Disease Journal* 24, no. 9 (2005): 755–60, doi:10.1097/01.inf.0000177279.72993.d5; G. L. Armstrong et al., "Childhood Hepatitis B Virus Infections in the United States before Hepatitis B Immunization," *Pediatrics* 108, no. 5 (2001): 1123–28, doi:10.1542/peds.108.5.1123.
3. CDC, "A Comprehensive Immunization Strategy."
4. Shepard et al., "Epidemiology of Hepatitis B."
5. J. Hayashi et al., "Hepatitis B Virus Transmission in Nursery Schools," *American Journal of Epidemiology* 125, no. 3 (1987): 492–98; E. D. Mcintosh et al., "Horizontal Transmission of Hepatitis B in a Children's Day-Care Centre: A Preventable Event," *Australian and New Zealand Journal of Public Health* 21, no. 7 (1997): 791–92, doi:10.1111/j.1467-842X.1997.tb01797.x.
6. S. F. Schillie and T. V. Murphy, "Seroprotection after Recombinant Hepatitis B Vaccination among Newborn Infants: A Review," *Vaccine* 31, no. 21 (2013): 2506–16, doi:10.1016/j.vaccine.2012.12.012.

7. Armstrong et al., "Childhood Hepatitis B Virus Infections."
8. CDC, "A Comprehensive Immunization Strategy."
9. Shepard et al., "Epidemiology of Hepatitis B."
10. Ibid.
11. CDC, "A Comprehensive Immunization Strategy."

C. DO WE GIVE TOO MANY VACCINES TOO SOON?

1. Centers for Disease Control and Prevention (CDC), "Immunization Schedules: Recommended Immunization Schedule for Persons Age 0 through 18 Years," 2014, http://www.cdc.gov/vaccines/schedules/hcp/imz/child-adolescent.html.
2. P. A. Offit et al., "Addressing Parents' Concerns: Do Multiple Vaccines Overwhelm or Weaken the Infant's Immune System?" *Pediatrics* 109, no. 1 (2002): 124–29, doi:10.1542/peds.109.1.124.
3. S. Otto et al., "General Non-Specific Morbidity Is Reduced after Vaccination within the Third Month of Life: The Greifswald Study," *Journal of Infection* 41, no. 2 (2000): 172–75, doi:10.1053/jinf.2000.0718.
4. F. DeStefano, C. S. Price, and E. S. Weintraub, "Increasing Exposure to Antibody-Stimulating Proteins and Polysaccharides in Vaccines Is Not Associated with Risk of Autism," *Journal of Pediatrics* 163, no. 2 (2013): 561–67.
5. M. J. Smith and C. R. Woods, "On-Time Vaccine Receipt in the First Year Does Not Adversely Affect Neuropsychological Outcomes," *Pediatrics* 125, no. 6 (2010): 1134–41, doi:10.1542/peds.2009–2489.
6. Institute of Medicine (IOM), *Childhood Immunization Schedule and Safety: Stakeholder Concerns, Scientific Evidence, and Future Studies* (Washington, DC: National Academies Press, 2013), http://www.iom.edu/Reports/2013/The-Childhood-Immunization-Schedule-and-Safety.aspx.
7. A. Rowhani-Rahbar et al., "Effect of Age on the Risk of Fever and Seizures Following Immunization with Measles-Containing Vaccines in Children," *JAMA Pediatrics* 167, no. 12 (2013): 1111–17, doi:10.1001/jamapediatrics.2013.2745.
8. D. S. Ramsay and M. Lewis, "Developmental Change in Infant Cortisol and Behavioral Response to Inoculation," *Child Development* 65, no. 5 (1994): 1491–502, doi:10.2307/1131513.

D. DO VACCINES CAUSE AUTISM?

1. A. Kennedy et al., "Confidence about Vaccines in the United States: Understanding Parents' Perceptions," *Health Affairs (Millwood)* 30, no. 6 (2011): 1151–59, doi:10.1377/hlthaff.2011.0396.
2. A. J. Wakefield et al., "Ileal-Lymphoid-Nodular Hyperplasia, Non-Specific Colitis, and Pervasive Developmental Disorder in Children," *Lancet* 351, no. 9103 (1998): 637–41.
3. D. K. Flaherty, "The Vaccine-Autism Connection: A Public Health Crisis Caused by Unethical Medical Practices and Fraudulent Science," *Annals of Pharmacotherapy* 45, no. 10 (2011): 1302–4, doi:10.1345/aph.1Q318.
4. Editors of the Lancet, "Retraction: Ileal-Lymphoid-Nodular Hyperplasia,

Non-Specific Colitis, and Pervasive Developmental Disorder in Children," *Lancet* 375, no. 9713 (2010): 445, doi:10.1016/S0140-6736(10)60175-4.

5. American Academy of Pediatrics (AAP), "Joint Statement of the American Academy of Pediatrics (AAP) and the United States Public Health Service (USPHS)," *Pediatrics* 104, no. 3 (1999): 568–69.
6. T. W. Clarkson and L. Magos, "The Toxicology of Mercury and Its Chemical Compounds," *Critical Reviews in Toxicology* 36, no. 8 (2006): 609–62, doi:10.1080/10408440600845619.
7. Institute of Medicine (IOM), *Immunization Safety Review: Vaccines and Autism* (Washington, DC: National Academies Press, 2004), http://www.iom.edu/Reports/2004/Immunization-Safety-Review-Vaccines-and-Autism.aspx; IOM, *Immunization Safety Review: Thimerosal-Containing Vaccines and Neurodevelopmental Disorders* (Washington, DC: National Academies Press, 2001), http://www.iom.edu/Reports/2001/Immunization-Safety-Review-Thimerosal-Containing-Vaccines-and-Neurodevelopmental-Disorders.aspx; V. Demicheli et al., "Vaccines for Measles, Mumps and Rubella in Children," *Cochrane Database of Systematic Reviews* 2 (2012): CD004407, doi:10.1002/14651858.CD004407.pub3; A. M. Hurley, M. Tadrous, and E. S. Miller, "Thimerosal-Containing Vaccines and Autism: A Review of Recent Epidemiologic Studies," *Journal of Pediatric Pharmacology and Therapeutics* 15, no. 3 (2010): 173.
8. L. E. Taylor, A. L. Swerdfeger, and G. D. Eslick, "Vaccines Are Not Associated with Autism: An Evidence-Based Meta-analysis of Case-Control and Cohort Studies," *Vaccine* 32, no. 29 (2014): 3623–29, doi:10.1016/j.vaccine.2014.04.085.
9. A. M. Persico and V. Napolioni, "Autism Genetics," *Behavioural Brain Research* 251 (2013): 95–112, doi:10.1016/j.bbr.2013.06.012.
10. R. Stoner et al., "Patches of Disorganization in the Neocortex of Children with Autism," *New England Journal of Medicine* 370, no. 13 (2014): 1209–19, doi:10.1056/NEJMoa1307491.
11. W. Jones and A. Klin, "Attention to Eyes Is Present but in Decline in 2–6-Month-Old Infants Later Diagnosed with Autism," *Nature* 504, no. 7480 (2013): 427–31, doi:10.1038/nature12715.

E. DO VACCINES INCREASE A BABY'S RISK OF SIDS?

1. M. M. Vennemann et al., "Do Immunisations Reduce the Risk for SIDS? A Meta-analysis," *Vaccine* 25, no. 26 (2007): 4875–79, doi:10.1016/j.vaccine.2007.02.077.

F. SHOULD WE WORRY ABOUT ALUMINUM IN VACCINES?

1. R. J. Mitkus et al., "Updated Aluminum Pharmacokinetics Following Infant Exposures through Diet and Vaccination," *Vaccine* 29, no. 51 (2011): 9538–43, doi:10.1016/j.vaccine.2011.09.124; H. Hogenesch, "Mechanism of Immunopotentiation and Safety of Aluminum Adjuvants," *Immunotherapies and Vaccines* 3 (2013): 406, doi:10.3389/fimmu.2012.00406; N. W. Baylor, W. Egan, and P.

Richman, "Aluminum Salts in Vaccines: US Perspective," *Vaccine* 20, suppl. 3 (2002): S18-23, doi:10.1016/S0264-410X(02)00166–4.

2. Mitkus et al., "Updated Aluminum Pharmacokinetics."
3. R. E. Flarend et al., "In Vivo Absorption of Aluminium-Containing Vaccine Adjuvants Using 26Al," *Vaccine* 15, no. 12–13 (1997): 1314–18, doi:10.1016/S0264-410X(97)00041–8.
4. Mitkus et al., "Updated Aluminum Pharmacokinetics."
5. P. A. Offit and C. A. Moser, "The Problem with Dr Bob's Alternative Vaccine Schedule," *Pediatrics* 123, no. 1 (2009): e164-69, doi:10.1542/peds.2008–2189.
6. Mitkus et al., "Updated Aluminum Pharmacokinetics."

G. HOW DO WE KNOW THAT BABIES NEED SO MUCH IRON?

1. Joint FAO/WHO Expert Consultation on Human Vitamin and Mineral Requirements and WHO Department of Nutrition for Health and Development, *Vitamin and Mineral Requirements in Human Nutrition*, 2nd ed. (Geneva: World Health Organization, 2005).
2. K. M. Hambidge et al., "Evaluation of Meat as a First Complementary Food for Breastfed Infants: Impact on Iron Intake," *Nutrition Reviews* 69 (2011): S57-63.
3. L. Davidsson et al., "Iron Bioavailability in Infants from an Infant Cereal Fortified with Ferric Pyrophosphate or Ferrous Fumarate," *American Journal of Clinical Nutrition* 71, no. 6 (2000): 1597–1602.
4. FAO/WHO, *Vitamin and Mineral Requirements.*
5. K. G. Dewey and C. M. Chaparro, "Session 4: Mineral Metabolism and Body Composition Iron Status of Breast-Fed Infants," *Proceedings of the Nutrition Society* 66 (2007): 412–22, doi:10.1017/S002966510700568X; U. M. Saarinen and M. A. Siimes, "Iron Absorption from Breast Milk, Cow's Milk, and Iron-Supplemented Formula: An Opportunistic Use of Changes in Total Body Iron Determined by Hemoglobin, Ferritin, and Body Weight in 132 Infants," *Pediatric Research* 13, no. 3 (1979): 143–47; M. Domellöf et al., "Iron Requirements of Infants and Toddlers," *Journal of Pediatric Gastroenterology and Nutrition* 58, no. 1 (2014): 119–29.
6. U. M. Saarinen and M. A. Siimes, "Iron Absorption from Infant Milk Formula and the Optimal Level of Iron Supplementation," *Acta Paediatrica* 66, no. 6 (1977): 719–22.

H. SHOULD WE BE CONCERNED ABOUT ARSENIC IN RICE CEREAL?

1. U.S. Food and Drug Administration, "Metals: Arsenic in Rice and Rice Products," accessed January 10, 2014, http://www.fda.gov/Food/FoodborneIllness Contaminants/Metals/ucm319870.htm.
2. U.S. Environmental Protection Agency, "Arsenic, Inorganic (CASRN 7440–38–2)," Integrated Risk Information System (IRIS), 1998, http://www.epa.gov/iris/subst/0278.htm.
3. World Health Organization, "Arsenic," International Programme on Chemical

Safety, accessed January 10, 2014, http://www.who.int/ipcs/assessment/public_health/arsenic/en.

4. Ibid.
5. National Research Council and Committee on Inorganic Arsenic, *Critical Aspects of EPA's IRIS Assessment of Inorganic Arsenic: Interim Report* (Washington, DC: National Academies Press, 2014), http://www.nap.edu/catalog.php?record_id=18594.

INDEX

Page numbers in italics refer to figures and tables.